The Fat Boy and The Money Bomb

A Story by

William C. Sailor

Gail –

Thanks for the support through the years!

Bill

A Money Bomb Press Book, Published by Lightning Source,
a subsidiary of Ingram Press Publishers

Money Bomb Press Books are available through Ingram
Press, and available for order through Ingram Press
Catalogues

Visit my website at www.moneybombpress.com

Printed in the United States of America

First Printing: June 2013
Money Bomb Press, LLC

ISBN: 978-1-62747-009-4
eISBN: 978-1-62747-010-0
LCN

Acknowledgements

This book was written with the encouragement, wisdom and zeal provided by Tom Bird and Ramajon at Sojourn Publishing LLC. Thank you to Dan Gerth for the classified content check. I also wish to thank my wife and partner Judith Horning Sailor for her patience, love and tolerance.

Foreward

This story is of a young man, Stanley Hall, who ends up changing "business as usual" at a nuclear weapons laboratory. It has a clear beginning middle and end, which I hope will make it easier for the reader to follow. The brevity of this book also should make it easy on the reader, but I hope that there is something that will live on in the mind of the reader after the book is finally put away.

The intent of the first third of the book is to engage the reader by sharing childhood experience. The young boy has some unfortunate circumstances in his family but is also fortunate to have a good set of friendships. Not all readers will relate when the childhood play evolves into juvenile delinquency and borderline criminality, but those tales should add amusement.

Loss of innocence is a theme in the second part of the book that many will understand. The innocence of the early life vanishes in a flash and the trials of navigating a path in the adult world come far too quickly for this boy who has suddenly become a young man.

In the last third of the book, after he has invested in a college education, his illusions about the world "out there" are harshly challenged. He feels compelled to make a decision. Does he play along with the corrupt elders that he works with, or should he risk losing his investment? He seeks answers in the intellectual world of modern game theory. He learns, as many of us have, that the intellect is only capable of providing a limited set of options.

The reader will want to know if this story is autobiographical. The answer is yes and no. The initial character description at the start of the book is indeed based on myself at that young age. Fortunately for the reader, the story and the main character become purely fictional after only a few pages.

Table of Contents

School is Out

The end of ninth grade for me happened on a Friday afternoon in June in Palo Alto, California. My name is Stanley Hall, but people just call me Stan.

Palo Alto is a small city south of San Francisco on the peninsula separating the San Francisco Bay from the ocean. At the top of the peninsula is San Francisco; at the bottom is San Jose. The area where Palo Alto is located is called Silicon Valley, because of all the semiconductor industry there. Back when I was a little boy, the integrated circuit was just in its infancy, and Palo Alto was simply known as a good place to live, and the location of Stanford University.

Indeed, it was a good suburban place to live. The streets were clean, potholes usually kept to a minimum, and houses uniformly middle class looking. On the north side of town were many of the older homes with Spanish tile roofs and road lined with maple, oak, and eucalyptus trees. This was the side of town nearest the university. When the school was first built in 1893, Leland Stanford had already owned the land and used it as an experimental farm for growing trees to make into railroad ties for his Southern Pacific Railroad. Hence, the abundance of eucalyptus trees in Palo Alto, which were imported from Australia as moneymaking venture.

The trees help to make the Stanford campus and the town beautiful. The campus is to the north and west of the town and keeps a somewhat separate existence from Palo Alto, possibly because of the original intent of

Leland Stanford, who wished to keep the students away from the vices that are available in college towns, such as alcohol. In fact, for the first several decades after the school had opened, Stanford had managed to pass a law banning the sale of alcohol in the Palo Alto city limits. The nearest location for alcohol purchase was across the county line in East Palo Alto.

The south side of town had been filled with fruit tree orchards for these several decades, and fruit canning was the main industry in San Jose for many years. The rows of apricot trees had been gradually cut down and filled in with houses over the years, especially under the direction of one J. D. Eichler, who had covered the south side with "affordable" wooden houses of a more modern design. These houses were either loved or hated by the town folk who just called them Eichlers. They were a classic set of tract homes from the 1950s, each occupying exactly the same size lot (1/8 acre), and each house having one of five different floor plans. The interior walls were stained wood originally, but over the years, some owners would paint the walls lighter colors to brighten the house.

Many houses still had the original carport in front, but many carports had been replaced with garages. Asphalt driveways were the originals, but replacement with concrete was also common. What most families had in common with each other was the need to not stand out and be too different. In other words, conformity was the norm. Looking a little better than the neighbors was okay, so long as nothing was too ostentatious.

To the bay side of the town was a light industrial area, having a few companies such as Ford Aerospace, Teledyne Industries, and Beckmann Instruments. Nearby cities, which had long ago ceased to be separated by any barren land, included Mountain View, Menlo Park, Sunnyvale. Each of these had bayside industrial areas that were small in land area compared to the two behemoths of the south bay: Moffett Field and Leslie Salt.

Moffett Field was an operational Naval Air station that was used to launch sea patrols by the P-3 Orion subchaser aircraft. These large propeller planes took off around the clock roughly every four hours and patrolled a large section of the ocean off the coast of California and Oregon for Russian submarines. There was often a drone of propeller noise over the area because of the planes. The strip had originally been built for the US Army Air Corps dirigible ("blimp") fleet. Hence, there were several large blimp hangars on the air base that were visible from most of the peninsula.

Next to the airfield is a large Christmas tree farm, accessible via the frontage road along Highway 101.

To the east of Palo Alto towards the bay were the salt ponds of the Leslie Salt Company, which were still operating and producing table salt. A salt pond is essentially a shallow piece of water that is blocked off by levees to protect it from the tides. The stagnant water, through the process of evaporation, becomes ever more concentrated in salt, until salt cakes form on top of the mud. The process takes at least one year for each salt pond, during which the pond becomes extremely, shall we say, pungent, with the smell of stagnant bay water. Adding to the interesting scenery along the bay are the seagulls ducks and cranes, which seem to greatly enjoy the stagnant water and do not mind the smell.

Separating Palo Alto from the baylands to the east is the six-lane US Highway 101 and its two frontage roads. Since this highway was built on top of mud, its pavement is not in the best repair and continually needs repaving. It is also one of the busiest highways in Northern California.

The roadway rises and dips, rises and dips because of the soft layers underneath the asphalt. When the rains start in the fall, the road can become stick, as it can on any highway. But on US 101, the dips in the asphalt will fill completely with water upon heavy rains, inviting the condition called hydroplane. Hydroplane is the result of the car tire completely losing contact with the asphalt because a layer of water occupies the space and does not move out of the way promptly enough. It was under those conditions during the middle of my ninth grade when my older brother, Bob, was killed returning home from his girlfriend's house. It was 4:00 AM on a Sunday morning and he completely lost control of the car, went off the side of the freeway, hit a telephone pole and was killed instantly.

The shock of the event hitting our household is still unforgettable. My mother was crying uncontrollably and her mother came over to try to comfort her. It was just the three of us in our Eichler house. The minister from our church, Pastor Frank, came over, but he did not seem to have anything to say, at least nothing that had any meaning to me. I looked at him and got angry and just walked out the front door. Then I got on my bicycle and rode away, not coming home until dinnertime, several hours later. When I got home, my mother and grandmother were still there and my mom was still crying. I just went to my room and closed the door. The next day, I just got up and went to school. The funeral service was on Friday and my mom asked me if I wanted to go and I just said no. I was thinking that I did not want to be around all that crying.

I needed to cry as much as anyone else, and going to that memorial service would have been the best thing I could have done. Perhaps some of the grief in the remainder of this story would have been avoided if I had made a better decision.

One may wonder where my father was in all this, but I never saw him. There was a place he used to go to drink in East Palo Alto called the Reno Club. I think that he was there the whole time. In any case, he did not live with us any more. I had not really seen too much of him since about third grade, the first time he was put in the hospital for drinking. I couldn't stand him.

My brother Bob, on the other hand, had been everything to me. We had gone through Little League together, him in the Major League and me in the minors. I was a pitcher and he was a catcher. In Palo Alto, the Major League was for those kids eight to twelve who were selected in the draft and made the cut to stay on the team. The minors were for those kids who just did not make the cut. No coach ever selected me for the majors, but Bob thought I was good enough. After Little League was over and I was thirteen, Bob coached a Junior League team (for those kids thirteen to sixteen), so I was picked to play in the "majors" at last. He was eighteen by then, and he had a car and drove me to practice and to the games. I was allowed to play in most of the games and even got a few hits.

That would be for only one season, because of his early death. I did not know what to think about the turn of events. I thought it was horrible how my mother was crying endlessly. It was worse than just simply losing my brother. I would try to make the best of things, and I simply went to school every day to get away from that sad home.

Fat Boy

By five o'clock on Friday the last day of school, I was already headed over to the Kozy house on my motorcycle. There were plenty of cars on the road, but I didn't see any cops. My motorcycle was a Yamaha 125, white, plain vanilla model. The muffler was quiet enough so that I didn't draw too much attention to myself. The only giveaway that I did not have a driver's license was that there was no license plate.

I had already delivered all the newspapers on my route that afternoon after school. I always used a bicycle for that, cloth shoulder sacks holding the papers; all delivered the same way every day, silently. At the apartment buildings, I just rode in, dropping papers one after the next, about a hundred a day. At the point where the route ended, there was a donut shop, and I could have a feast of a few jellyrolls.

The air rustled through my helmet and I kept checking the rearview mirror in case I was being followed. There were no problems so far. Come to think of it, I must have been pretty obvious spot as an underage unlicensed motorist, having no license plate. Very few people rode such a small motorcycle around Palo Alto to begin with. But my helmet was painted orange with yellow stripes, a design that I had made myself. Only a kid would do that.

My mom had bought me some black leather-riding boots for Christmas that year before, and they were a normal part of my getup. It had been understood that these were for off-road use only, and I has assured her that they were. Blue jeans sufficed for riding pants; real leather ones were too expensive and probably too hot for the summer.

I rode down Clarke lane with the throttle almost completely closed to keep the noise down. It was about 5:30 in the evening. Five houses before Clarke t-boned into Mendoza Avenue, I killed the engine and coasted. Silently approaching the Kozy house, my ears were treated to the rush of the wind through my helmet.

The bike dipped down into the gutter and back up on the sidewalk, then up the Kozy driveway. Kickstand pushed out, I leaned it over, kicked my body off the bike. The gloves came off next, then the helmet. The garage door was open a crack, so I headed in there rather than through the front door.

Inside was Bruce Kozy, the oldest son, working on what seemed like a brand new Ducati motorcycle. Ducati was one of the most elegant brand names of motorcycles in the world, and this bike was an exceptional example of the brand. The engine was a narrow V-Twin two-cylinder 750cc air-cooled motor, and the bike was made for very high speeds, having a full fairing to reduce wind resistance.

Bruce's Ducati was set up with a red gas tank and matching bright red fenders, probably made from fiberglass. It also had red side panels that covered the polished aluminum motor. The huge disk brakes were a prominent feature of the bike, and the wheels were polished magnesium. The tires on this motorcycle seemed new.

"Where did you get that?" I asked.

"Shut up Fat Boy," was the quick answer. He seemed quite dedicated to what he was doing. I stood back a few steps looking at him, he who was masterfully loosening the gas tank from the frame. There was no conversation, despite my attempts to start one.

"How fast have you gone on that thing yet?"

"Fast enough," was his reply.

Bruce always made me feel unwelcome, and this time was no exception.

His Corvette occupied the right half of the garage, and the thing was torn apart and was all over the place. I had never seen it run. The engine was tucked into a corner of the garage and the four fenders were in the backyard. The tires were still on the car so it could be moved as needed. The car was very light. Not so light that you could pick up a corner of it, but very light compared to a typical car. It has been raced years ago at Sears Point raceway by Bruce's uncle, Mitch. Next to the Corvette were about a dozen cardboard boxes of a variety of sizes.

The left half of the garage was full of motorcycles. There was an old Suzuki 500 that Mark had been riding on the salt levees for years. There were at least three nearly new street bikes, all of which ran and all were Bruce's. On the other side of the row of Suzukis was Bart's Kawasaki 250 Bombardier.

There was no sign of any of the brothers except Bruce, so I headed into the main portion of the house through the kitchen door.

The Family Next Door

The Kozy family was the family that I had lived next door to with my mom, dad and brother when I was a little boy. The Kozy family consisted of four boys, a mom and a dad. Their actual name was Kozumplick, but no one called them that. It was remarkable that they had four boys spaced five years apart. In second grade, I remember Bart was in the same grade with me, and Kurt was one year younger in first grade. Mark was in fourth and Bruce in fifth.

I was a best friend with Bart in second grade, and we used to ride bicycles together at lunchtime over an obstacle course we had made. Our teacher, Miss Miskell, thought that Bart was troublesome and used to pick on him for 'not using his brains" or something like that. I used to help him with his reading assignments. He used to say he just could not wait to get out of school to go home and work on his car with his brothers. That got me curious.

The first time I went over to the Kozy house, there was an old car in the backyard, a Desoto or something. The dad, George, had brought the thing over for the kids to play with. George and the mom, Midge, were divorced already, but George still came around to be with the boys sometimes.

George was the owner of a machine shop in Sunnyvale, one with only two employees. I think one of them was his girlfriend. He was just a really nice guy as far as I knew, and I liked the mom, too. I had no idea

11

why they had gotten a divorce, and I don't think the Kozy kids knew either.

There were always some other grown men around, because the mom, Midge, had three brothers who were considered part of the family. There was Uncle Mitch, who lived in Oakland across the bay, Uncle Milt, who lived in Texas but visited often, and Uncle Mack lived in Southern California.

As I said, they had fully enclosed yard and a wrecked car inside of it. I think the first car I saw there was an old Desoto or something, and I saw Bart had a screwdriver in his hand taking something out of the dash. That was "his car."

The two older boys started using power tools first. I saw Mark using an electric drill and I think Bruce was using a grinder the first time I came over to the house, probably when I was about seven years old.

Things were not going too well in my family the time. I hardly ever saw my dad—he was always at work. When he got home, Mom and Dad would fight. She said he was alcoholic, and I do remember him being pretty crazy. But most of all, he was incredibly fat, and I had no use for him.

There was one time I remember when my father had done his usual thing in the evening after work. He had come home in his suit from work and sat down at the dinner table like usual, drinking wine and eating his steak with his bare hands and getting steak sauce all over his face. Mom said he should use a napkin to wipe himself and he blew up – yelling—something about how she was always bothering him. He slammed down his fist on the table, but his fist caught the corner of his plate, so the plate just flew up in the air, with all his food going all over the place. He screamed at my mother, "Look at what you made me do!"

I was probably in first grade at the time and just started crying. At least I must have been because I remember him screaming at me, "Stop crying now!" Then he hit me in the face with the back of his hand and said, "Stop crying or I will hit you again." I stopped crying through the force of willpower at a very young age on that night. My older brother was just an observer that night, but on several other nights, he was the recipient of the rage.

I remember Dad leaving the dinner table and saying that he is going back to work right then. He got to the front door and my mother said clearly and loudly, "Bob, you are too drunk to drive." It was so weird to me at the time. He was still in his suit and was planning an evening in the office catching up on his work. I would expect her to say "It's too late to work," or even "You are too drunk to show up to work" or something like that. But he was apparently too drunk to even drive. He must have been very, very drunk. After a while, Dad moved out.

I liked going next door—it was more fun over there.

It was about seventh grade that we moved to another "Eichler" neighborhood in Palo Alto. I had finished sixth grade and started at Westwood Junior High School, meeting up with my old friend, Bart. Bruce rode a motorcycle to school every day. Mark wore a leather jacket to school, but didn't get a motorcycle until later.

Bruce was almost six feet tall in ninth grade, and was a star on the basketball team. Not only that, but he was a star on the football team as well, and popular with the girls. I suppose he was good-looking because he was voted "best looking boy" when he was in ninth grade. Bruce and my older brother Bob had some things in common, for sure. They both were good in sports and both were popular with the girls. Although my brother was a year older than Bruce, it was Bruce rather than Bob who was generally first string at every sport.

The younger Kozy brothers were also slender and athletic, but none of them played in school sports—they just rode motorcycles. Mark was almost as tall as Bruce, but even thinner to the point of being downright skinny. He said he tried various ways to put on weight, but nothing seemed to work. He was always dressed in Levi's pants and jacket. In later years, he always had a motorcycle helmet with him.

Bart, the next youngest brother, was probably half a foot shorter, and he virtually worshipped the two older ones. He wore a leather motorcycle jacket everywhere even while he was just on a bicycle. I remember seeing him when he was only about twelve years old, he was able to ride a bicycle on its back wheel for a block at a time, like it was a unicycle. It was really phenomenal to watch, as I had seen "wheelies' before, but a wheelies usually just lasts a few seconds. Bart could ride the wheelie for minutes at a time. One day, something went wrong, however, and he fell and landed with his knee on the concrete, giving him a limp that he still has today. But that limp did not stop him from racing motorcycles by the time he was sixteen.

The youngest brother, Kurt, seemed to be a physical replica of Mark, except a few years younger. They had exactly the same face, and they both looked a lot like their dad, George. Kurt also was exceptionally skinny. Unlike Mark, who was aloof and a bit domineering, Kurt was very friendly and accessible. Kurt was Bart's constant companion.

Evening Meeting

I entered the main part of the Kozy house, after leaving Bruce with his Ducati. Mark was sitting at the kitchen table with his mom, Midge, eating something that looked like soup and talking intently with her about the bible. Midge said, "Most people think that there are only ten commandments for us to follow, as Moses said. But there are many more commandments than that."

Mark said, "Like what?"

"There is the command to be loyal to your family. That's everywhere in the Old Testament."

"It is, huh," said Mark, less enthusiastically.

I sat down and listened, because to me it sounded like an interesting conversation. Midge continued, "God wants you to treat yourself well and not let yourself get tempted by drug abuse. Using LSD is a pathway for Satan to enter your soul."

Mark said "you sure know a lot, Mom," again with a slight hint of sarcasm.

It was time for me to interrupt, "What does God think about war? Does God think war is good?"

Midge said, "Only when you are on the side of the Lord. Men who are in combat must pray and confess their sins before going into battle. That is what God wants. Furthermore, the war must be on the side of a nation that is Christian against one that is not. There is no such thing as a just war that is fought by the UN or NATO or anything like that."

I thought that was really interesting, and I had gone to church for years and never heard anything like that. But I had enough. "Where's Bart?" I asked.

"He's in his bedroom," said Mark. As I walked away, Midge said, "Ask him if he wants dinner."

The door was closed, so I opened it and walked in. Bart was sitting with Kurt on the floor, inspecting a small pistol. It was partially taken apart and the ammunition clip was set over to the side. "Hey there, Stan," Bart said in his usual friendly voice. "We are having some problems with misfires with this thing."

"What is that?" I asked.

"It's one of Uncle Mitch's extra pistols. He gave it to me."

Then he continued, "It is a .22 automatic with a nine-shot clip. You can hide the thing in the pocket of your pants and have it out in your hand in no time. There is just a problem with the firing pin, I think." He intently was studying the firing mechanism. "Maybe the pin is worn down."

I noticed a smell of gunpowder in the room as if he had already been firing the thing. Indeed he had. There was a pile of phone books in the corner of the room with bullet holes in them. "Bart, let me go look at that," I said while I went over to the phone books. Indeed, there were bullets embedded a few inches inside. I opened up one of books; it was the yellow pages, starting at A. For about A through G, the bullet made a nice clean little hole. By about M, there was an awkward kind of rip. When you got to S, where the bullet stopped, the pages were kind of shredded and the bullet had deflected sideways. It was interesting.

Midge came. "Bart, did you want any dinner?"

"No, Mom." She departed and Mark came in.

"Are you still screwing around with that old piece of shit?"

To that, Bart replied, "I need to weld an extension to the firing pin. It is worn down too much."

"That was just an old piece of Italian junk to begin with," retorted Mark.

Then Bart said, "Don't worry, I'll get Bruce to help me with this."

Then Mark changed the subject. "I just got talking to Bruce and he wants to go over to the Stanford pool tonight."

Bart got interested. "Yeah, when?"

"Sometime after dark, probably about nine."

Everyone in the room agreed it would be a great thing to do.

Stanford Pool

The lights were off at the pool and we probably did not get there until 10:30 at least. All the students were studying or something because the place was silent. We moved forward in the car and I kept staring out front in the back of the backseat to see what's going on. As the fence approached, Bruce switched off the headlights. We drove around to the parking lot near the entrance to the locker room. Kurt got out. Kurt opened the rear of the station wagon from inside and got out with a bicycle. All of us, including Bart and Mark, got out then. We closed the car doors as quietly as possible then moved to the fence.

Bruce climbed the fence first with Kurt right behind. Bruce hooked the handlebar through the links as the bike made its way up the fence. The fence must have been 12 feet high. At the top, Bruce reached down. He deftly carried the bike from one side of the fence to the other, hooking the handlebar on its way down, skillfully lowering himself from the inside of the barrier that was built to keep us out. We all scampered over.

The shoes came off. They were only needed to climb the fence. When we were all in the water, Kurt bounced off the diving board. I think he was a bit of a showoff. Bart had the bike and circled the Olympic size pool, getting up remarkably good speed for going in. As he is going off the edge of the pool, the front wheel caught the water and the bike overturned completely, and his face solidly hit the water. Laughter. "I'm next" I let everyone know and I swam to the bottom of the pool, grabbed the wheel (The bike was upside down on the bottom of the pool resting on its seat and handlebars), and I swam to the top surface to haul it out.

There is no way that Bart is going to get all the attention here, I may have been thinking unconsciously (In fact, I don't remember what I was thinking at all.). I just did it—I rode the bike off the diving board. The bike is difficult to balance up there, about 3 feet off the ground. I told myself, "Careful. You don't want to fall off the side. You will look like a fat fool." The others stopped what they're doing and watched.

It is very hard to get enough speed to even balance on the length of the diving board, let alone fly off the end of the board. It would be a disaster—total humiliation—to fall off the side of the diving board. So I pushed myself forward with one single kick of the pedal and was at the edge of the diving board instantly. The sprocket caught on the edge and made a loud sound and the chain wheel was forced forward. The front of the bike and the handle bars disappeared from view and my face was in the water before I even knew what was going on. Underwater, the bike dropped below me and out of sight. After paddling to the surface and gasping for air, there was no way to stop laughing. "Awesome!" The yelling was coming in. Applause. This was an award-winning moment for me with my peers. I got a little afraid because we were making so much noise. "Quiet you guys," I shouted.

Bruce showed no expression. "Get out of my way!" he said, and dove into the pool right to the bottom where the bike was. In a manly, athletic way, he grabbed the thing and had it out of the water in no time. Quickly moving it over to the next diving board, he did what no one had even talked about doing. He started climbing the stairs leading to the Olympic high dive platform. This proved a bit of a challenge carrying the bike, so he told Kurt to help him. The balancing act on the high dive was a bit scary to me. It seems like 20 feet to the concrete below, but he just went ahead and did it. As he rode to the edge of the diving board, predictably the chain wheel caught on the edge and he wildly flew over the handlebars, with the bike shooting below him right straight down into the water. He tried to push himself into a position so he will not hit either the bike or the water too hard, and he salvaged some sort of dive out of the whole thing. He didn't look too bad.

Bruce drove again on the way home. He was the only one of us who actually had a driver's license then. It was summer and the air was warm, so we weren't too uncomfortable.

He said to the rest of us, "Kids in the back."

I think that he meant anyone without a driver's license was a kid. Mark had his learner's permit so he rode up in front in the passenger seat.

"Kids in the back," he echoed.

"You really fucked up on the low dive," said Bruce. "You almost broke the chain wheel on my bike."

"The bike is a piece of shit," said Mark.

This conversation was not going to go too far. Then Bruce said, "Hey, you guys have to fill up my car with gas after I drove you over here."

Then Bart volunteered, "I know there is a gas station where we can fill up the car over on Middlefield Road." This came out of the very back of the car.

When we arrived at the Kozy house, we all wanted out of the car right away. The garage served as a good changing room for me, who was not officially a member of the family.

Bart finished dressing first and went to get the bolt cutters. They looked like the biggest pair of pliers I'd ever seen, with probably 3-foot long handles.

"What the heck are those?" I asked while tying my shoes.

"The Shell station is closed. This is how you get gas. We will just cut the lock. I saw Uncle Milt do this once in Texas."

Mark had gotten the car keys from Bruce. This trip had to be planned out much more carefully than the trip to the Stanford pool, and Mark was clearly in charge. "I am not putting any Shell gas in any car—that shit sucks. Shell? You've got to be kidding."

"Oh, I suppose you want Union," mocked the younger brother Bart, who only was fourteen.

"The Union station is right down the street, spook." So Mark drove and the rest of us went along to get some gas. Bruce stayed home.

"Go past it once," said Bart, giving orders from the backseat.

"Shut up, Spook," from Mark.

That was one of Bartholomew's other nicknames when he was getting a bit out of place. On the list of available names was also the N-word, "Mexican," and "Gook" (All the kids were third generation Polish, but I never once heard them call each other "Polack.").

It was a busy intersection during the day at the corner of two four-lane roads. There were some large satellite dishes in the parking lot across the street at Ford Aerospace. Down Middlefield road was Fairchild Instruments, where the first integrated circuit was made. But no one cared. What I cared about was getting gas for my motorcycle. I was sure that I was entitled to some, being part of this operation. We circled the block and parked. The plan was as follows: Bart would cut the lock off with the bolt cutter. We would park at the pumps furthest away from the main intersection so no one would notice.

While we were across the street planning, Mark yelled "Cops!" suddenly as a car went past. Huge adrenaline rush. It was not a cop car. "Okay, are we ready?"

We pulled ahead, adrenaline surging, into the gas station. It must have been 12:30 in the morning. The bolt cutters were skillfully applied to the lock and the pump nozzle put into the car. Bart squeezed the handle. Nothing.

"Spook," said Mark, "turn on the handle."

Bart complied. Surely gas would be coming out now, but there was no difference.

"Hit it—the pumps are turned off."

We all flew back into the car and slammed the doors. We were down Middlefield going probably 70 miles an hour. "Mark, slow down," said Bart. He did. We drove the speed limit back home and called it a night.

Earning a Promotion

Rudy was another family/gang member who had been brought in as a loner. He was the older of two kids—his younger sister was a party girl. At age fourteen, he already had a Chevy Biscayne. It had no license plates and it barely ran.

Rudy and I shared a hobby and that was rabbit hunting. The hunts were just outside of town in the wetlands along US Highway 101, and we regularly drove the Biscayne on the hunts. The wetlands, or "baylands" as we called them, were rich in these large rodents. A single trip could net two or three, which we would simply put in the trunk and take home to Rudy's house for storage in the family freezer. Someday, we would learn how to gut and clean them. There was another guy at the junior high who knew how, and he would teach us.

As time went by, the freezer got fuller with the carcasses. Rudy got in trouble with his mom and dad for that, eventually. But it took a while for them to notice. They were busy, being surgeons at Stanford Hospital.

One evening (using spotlights to enhance the hunt efficiency) on the frontage road, the Biscayne quit in a bad way. The engine ran out of coolant completely and overheated, letting out a lot of steam and smoke. We were stranded just outside the city limits, in an industrial park.

Rudy let out the usual set of cuss-words—there was nothing new in that. Walking home was new. We locked the guns in the trunk of the car with the night's harvest.

The closest house we could walk to for help was the Kozy house. After an hour walk, we found Bart at home working on his motorcycle. He looked up.

"The Biscayne quit," said Rudy.

No one was interested in walking back out there and getting the thing running. We waited until Bruce got home and talked him into driving us out there in his Chevy Impala Wagon with some tools.

"Get your stupid ass in the car," said Bruce. That was about as friendly as he ever got.

Out on the edge of town, past Ford Aerospace, next to the baylands and the salt levees, we spent probably an hour working on the car. It started out as a repair operation, but ended up as a salvage operation. The engine was frozen, meaning that the crankshaft would not turn. It would never run again. The battery, the guns, the bullet supply, the rabbits, and a few other items were retrieved.

As we were preparing to depart, Rudy suddenly put his .22 caliber rifle to his shoulder and shot the windshield of the car.

"God damn stupid shit!" said Bruce.

Rudy then delivered at least fifteen shots in rapid succession to the left side of the Biscayne, puncturing all the body panels from the rear fender to the front. Bruce was not too happy with this defiance, and shot Rudy full of verbal bullet holes, so to speak. The threat

Bruce made, that we would all have to walk home unless this stopped, was adequate to get Rudy into the Impala, and all of us drove home.

Back at the Kozy house, Rudy insisted that the Biscayne was evil and needed to be punished. Bruce wanted no part of this discussion and departed the house. Mark had come home and was changing the oil in his 57 Chevy Pickup. He was also performing some sort of tuning operation on the engine.

A fire extinguisher was available, obtained by the Kozys from a nearby apartment complex, and it was being used to store used motor oil. This was the type of fire extinguisher where the top could be removed by unscrewing. Normally, a gasket maintained the pressure seal. When there was no pressure in the tank, the top could be unscrewed and any liquid put in. There was an inlet valve that could be used to repressurize the unit at any gas station.

The five quarts of used engine oil that Mark had just produced were introduced into the unit, along with another five quarts of gasoline. Full of the thrill, we got in the back of Mark's pickup truck with the fire extinguisher and drove to the Shell station to fill the unit with air. Bart and Rudy both hesitated.

"That's a bomb," said Bart.

I thought it was simply a great idea.

Out on the edge of town sat the Biscayne, already shot full of bullet holes. The seats were soaked in gasoline and the old car set on fire. Rudy was yelling obscenities at the thing. He introduced a few dozen more bullet holes into the burning car. Then I brought out our flamethrower.

As soon as I had it out of the truck, Mark said, "Stan, I'm getting away from here," and ran a couple hundred feet down the road. Bart and Rudy went with Mark. My moment as a soloist had appeared.

I walked up to the Biscayne and emptied the contents of the fire extinguisher in a glorious burst of white-hot flames, streaming wonderfully from the nozzle of the (former) piece of safety equipment. I remember the skin on my face feeling kind of hot that night. I think I had to take a few steps backward, but that did not significantly diminish the bravado of the act in the eyes of my friends.

The political impact of the evening was large. Bruce removed himself from the gang as of the next day. While Mark and Bart became the chief executives, I became a full member, one who would help plan and perform dangerous criminal activities. That was all, of course, in the name of having fun.

Religious Training

During that week, the only way I could ride my motorcycle was to ask my mom for money. I would not be collecting any money for my paper route for another three weeks at least. All my money had been spent on a new rear tire. I promised my mom that I would not ride the motorcycle on the street if she just bought me some gas. I got the lawnmower gas can and carried it on my bicycle about a mile down the road to the Shell station to fill it up. Then I rode back with the gas can balanced on the handlebars.

With the motorcycle full of gas, I got my helmet, gloves and boots, and pushed the Yamaha down the street away from my mom's house down past the corner. After one more block, I turned left at the corner to the entrance to the elementary school. Helmet on, bike started, I rode across the lawn and through the buildings—the teachers were still there but little kids (Little kids: ones without motorcycles) gone. I went out the back of the school onto the lawn of Mitchell Park across the creek on the pedestrian bridge into the back of the Unitarian church, where there was a hole in the fence. Then through the church, the shopping center back alley, across the high school parking lot, and onto Mendoza Avenue. I went down Mendoza to the Kozy house—the headquarters for the planning of crimes, for motorcycle repairs, and for religious training.

Bart was there, Mark and Bruce gone. Ma was blending some veggie drinks for herself. "God wants you to be healthy, would you like some vegetable juice?" she asked when she saw me at the door.

She had the blender going with the top off and a dark green liquid filled the lower half of the blender bowl. "It's very delicious," she insisted.

"No, thanks," I said.

"Are you sure?" Then she picked up a cup of sugar. "I use all natural ingredients," she said while she poured the cupful of sugar into the liquid. The blender slowed slightly and I thought to myself, *How weird . . . that drink is half sugar.*

She said, "Satan did that" while pointing at my stomach. "He made you fat. Satan wants you to eat bad food and be unhealthy and turn away from God." She said this in a perfectly confident, straight face. She was as skinny as a pencil, so to speak. "God gave us natural food. God is what you need to be healthy." I think this was some sort of challenge—at least it felt that way. "Bruce is skinny, but he doesn't believe in God," I said. "But I do." This was a ninth grade boy's attempt at *reductio ad absurdum*, which just took her on a new track.

"That doesn't matter" she said, "because Bruce accepted Jesus as his Savior when he was a little boy. God has forgiven him for being wrong. Once you believe in God and accept Jesus, you are saved. It cannot be undone even if you want it to be, even if you completely renounce God."

I was thinking about how Bart, Rudy, and I were stealing hundreds of pounds of scrap aluminum every week from businesses in the industrial park—and how stealing was against the word of God. This was a relief. I was saved anyway.

Ma got busy with some of her natural products while I snuck out of the kitchen to see Bart. "Going riding?" I asked.

A Motorcycle Ride

I never once saw Bart do any homework or studying. I didn't know how often he even went to school. But we sure did like to ride motorcycles. Kurt was coming along and we would be out to the baylands by five. The best way out was right through Ford Aerospace, right past the antennas. There was a creek that went along the back of the parking lot and then went below the freeway.

We simply rode in the bottom of the creek, keeping full throttle the whole time so as to not get stuck in the mud. The creek went underneath the freeway and into the baylands to the edge of the dump. The dump had a fence around it and there was no reason to go there anyway. We just went around it, onto the levees in the back. The levees segregated the water into salt ponds. We followed the levees for miles, sometimes cutting fences when they are in the way. The goal was to reach Moffett Field (also known as Moffett Naval Air Station), where the good stuff was.

Kurt rode on the back of Bart's Kawasaki, and Rudy rode on the back of my Yamaha. We had all been to Moffett before. Our anticipation was high: a diesel tractor had been left unattended by the Navy. Night settled as we approached the tractor. So we took out our flashlights. "It is a CAT diesel engine," said Bart, keeping his voice to a whisper. That whispering didn't make too much sense, come to think of it. After all, the bikes we were riding had no mufflers and we could not even hear each other until the engine was shut off. Bart wanted to take the entire Caterpillar engine home.

We sat there on the cat tracks, the four of us, just planning.

"I want to start a business pumping mud. Dredging is what they call it. Uncle Milt does it in Texas and he is really rich." Bart was as serious as the Sermon on the Mount. "I want to get this engine home and put a cutting blade on the front. I'll build a boat to carry the whole thing."

"How are you going to build a boat?" I asked.

"It's easy, we have lots of wood at the house."

"Aren't you afraid it's going to get a leak?"

"No, Uncle Milt does this all the time. He buys the sealer army surplus—it's no problem. Now, let me tell you Stan, you can really make some money with this thing. You wouldn't believe how rich Uncle Milt is. He has a big Cadillac and a big house in Texas with the ranch and horses and everything."

We didn't quite know how to get the Caterpillar home though. We would need a pickup truck at least. No one except Mark had a truck and he already said he wouldn't help us. We had to find another truck.

"I know," I said. "I know where there is a truck. There's one parked behind the Shell station. It's just missing a battery."

"Yeah," Bart said, "I've seen that one. It's an old Studebaker—all we need is a battery. I can hotwire it."

The plan solidified. We would steal a battery then put it in the Studebaker to drive it out to Moffett Field on the levees to steal the Caterpillar engine, to put it in a boat we would build, and start a dredging business and make it rich. With the plan in place, it was time to go home.

We were restless with how late it had gotten, so we needed to make some shortcuts to get home a little quicker. The shortest way home from Moffett Field was through the Christmas tree farm then down Shoreline Boulevard.

Riding in tandem fashion with flashlights tied to our handlebars on both motorcycles, we had a very fine cruise through the orchards past the drive-in theater. There was not a car in sight as we pulled onto Sterling Avenue. Then there was a cop.

"We are busted!" yelled Rudy.

The front brake on my Yamaha got a serious test. With Rudy on the back, I damn near locked up the front wheel, while the rear wheel skidded. There was a quick U-turn, followed by a downshift. Back through the Christmas trees, heart pounding, we rode as fast as possible. As I write this all these years later, it seems a wonder that there were no fatalities that night. After twenty minutes of riding full speed, Rudy and I stopped and turned around. Bart and Kurt were nowhere.

The Perfect Crime

I didn't see Bart or Kurt for a few days but, lo and behold, a story got out around the school that the two of them had been arrested. They had not been seen at school either. I made my usual ride on my Yamaha over to the Kozy house, finding him working to get a battery out of his mother's car.

"Mom is away visiting her brother Mack—now is the time." Bart said. "I have to tell you Stan, those cops could not break me. They sat Kurt and me down and told us that if we didn't give the names of the other two kids we were riding with, we would be locked up in jail for a week. But me and Kurt would not break—no way."

I was aghast. They indeed had been arrested.

"They kept us in the police station all night. In the morning, they called Ma. We didn't get home until 9:00 AM."

We were too busy with the next adventure to spend too much time recalling the highlights of the last one. The Studebaker was left unattended after 8:00 PM when the gas station closed. We walked there in the afternoon to scope it all out, pretending to need gas for our motorcycles, taking a gas can with us. It all looked good. I remember getting close to the truck and noticing all the rust. I had never seen a vehicle where all the windows were broken. The hood was propped up and the battery was indeed missing. There were wires hanging out everywhere and the truck kind of smelled bad. But there was air in the tires.

We were at the gas station at 8:00 with Ma's battery to watch them close up. How long could it possibly take? We sat down on the sidewalk behind a parallel-parked car and peered over the hood. Lights out. Door closed. Car starts up, drives away. The four of us crossed the street. I carried the battery.

Bart had assured us he knew how to hotwire an old Studebaker. He was right about that. He didn't tell us it would take over an hour. After a while, we had the starter motor turning, but no other sign of life from the engine. We had to go home and get the full can of gas. It was pure luck that we had one available.

When the truck finally started, it was the most incredible rush of adrenaline I had ever had until that point—we were all yelling and shaking each other's hands. By midnight, we had the truck in the Kozy driveway.

It needed a little water in the radiator. The taillights didn't work but who cared—we were going to get the Caterpillar, no matter what. Bart had already put everything we needed in the driveway, including a full set of tools and a floor jack or two. And a little gadget called a come-along, kind of like a little handheld winch. The idea was to loosen the diesel off its mounting bolts, back this truck up in front of the tractor, and slide the engine forward using the come-along into the bed of the truck. I was a little leery of the plan, but Bart was confident. Rudy and Kurt would be there for some extra hands in case the engine was too heavy.

We drove the "Stu" to Moffett Field, with Kurt and Rudy in the back of the bed. Bart sat in the front seat and informed me about his adventure with the police.

"They really thought they could break us? They were just stupid. But, you know, they were really mean, I'll tell you. They were calling me names and everything."

"What names" I asked Bart.

"Four eyes." Bart wore glasses.

"Is that all?"

"He called me a candy ass, and a wimp, and all sorts of stuff like that."

Silence.

"Bart, do you believe in God? You know, what your mom says. God wants you to be good and obey the Ten Commandments and all."

"Sure," he said. "Jesus died on the cross for our sins."

"Yeah, but if you believe in God, why are we stealing this Caterpillar engine?"

I up-shifted the Studebaker. I'll never forget how that thing had a four-speed "crash box" transmission. Modern vehicles all have automatic transmissions, or if they have a manual transmission, they are designed with special gears called "syncros" that prevent grinding. In the Studebaker, which must have been forty-five years old back then, the gears ground like hell whenever you tried to shift. It could be embarrassing, under some circumstances.

Bart answered, "The military doesn't need this tractor, that's why it's out behind there."

That made sense. They don't need it, but we do. So it's not against the Ten Commandments. God won't be mad.

"Besides," Bart said, "It's all taxpayer money anyway. The government is stealing it from us."

I nodded. That made even more sense. I understood at that exact moment all that I needed to know to go ahead with the crime. And besides, I thought, I'm still a juvenile in case I get arrested. They'll just let me go.

The truck ran pretty smoothly out beyond the orchards. A fence had to be removed to proceed further. The bolt cutters worked well on barbed wire, too, and we just drove right over a fence post that was in the way. Approaching Moffett, we got a really good rush looking at the silhouette of the tractor against the lights of the Naval Air Station. I spun Stu into a wide turn and backed up to the front of the tractor. The first thing that was apparent was that the truck was about 10 inches too high in the air to ever slide the engine onto the bed.

After we paused for a while, Rudy finally said, "We will just raise the tractor by jacking it up." Then he got the jack out and slid it under the tractor.

"You idiot," said Kurt, "that's not going to work."

Rudy tried it anyway. As he pumped the jack handle, the base of the Jack slid further and further into the mud. When the jack dropped 6 inches into the mud, he finally gave up.

"I'm a fool," he said.

Bart was silently planning all during this time. He finally said, "There is no way this thing is going to work. We have just wasted all this time for nothing."

I stood there trying to figure some sort of way out. There was none. Bart convinced everyone that while we couldn't possibly get the diesel home, we could at least get the turbocharger off the engine. After all, it will make a car go fast, in case one of us ever got a car.

Once the Studebaker took us back to the Kozy house, we put it inside the garage in case the cops were looking for it. And there it stayed for a few weeks. The battery was returned to Ma's car in time for her arrival back from her brother's house. But what could we do about this truck? There was talk of trying to sell it. We would not get too much money, and we did not have a pink slip. We could just drive it somewhere and leave it. But it was a perfectly good truck. I really liked the way it had four speeds, even though the gears were noisy.

Mark was walking through the garage and offered some advice. Mark, the second oldest son, was working for his Uncle Mitch at the speed shop.

"Who in hell brought this piece of shit home?" he said, looking at Stu. "It's nothing but a bucket of rust," walking away.

Mark had already saved up enough money and had bought himself the Chevy pickup and a new street bike. Bart and I could not understand what it was that made him so mean-spirited when it came to Stu.

"It's just as good as any Chevy," said Bart.

"Get rid of it," said Mark.

Then I said, "Mark I want you to grind off the numbers from the truck and punch new ones."

I was referring to the Vehicle Identification Number that is unique to each vehicle. There is one on the body and one on the frame.

"That's not how you do it," he said. "You weld over the old numbers then grind the weld down. Then your reweld the spot again and grind it down again. Then you punch new numbers."

Either Bruce or Uncle Mitch had apparently shown him how to do this. I traded my half of the turbocharger for his performing the trick.

The truck sat in the garage for another week or so while the operation was performed. Then I got enough money from my paper route to buy a battery, and after asking Bart, I simply drove it home and parked it in my mom's driveway.

"What is that thing?" she asked.

"It's a Studebaker."

"Where did you buy it?"

"I just found it in a field. It just started right up when I put a battery in it."

"Just a minute, you don't even have a driver's license yet."

"That's okay, Mom, I'll be getting my learner's permit soon. You can teach me how to drive in this thing."

"Okay, okay…" she said, fading off.

Montana

Word came to me through Bart a few months later that we would not be seeing Bruce again for a while. He had gotten the idea that if you steal a motorcycle, repainted, and changed the numbers, no one would ever know you stole it. He succeeded with his first one. So he did it again and again. The process worked quite well and he made good money selling the bikes. The only problem was that he started bragging too much about what he was doing, and the cops closed in on him. Indeed he was gone for six months, to someplace in Montana.

I really didn't miss him though. He was kind of an asshole—kind of scary.

Bruce was just under eighteen when they put them away. He had always been the first to do everything. He was first with a go-kart, first with a motorcycle, first to get a car, first to steal a motorcycle, and so on. He really hated cats for some reason. He used to go out in his Chevy and then try to run them over whenever he saw them. Later, he would take a gun out with him in the car so he could just shoot them if he couldn't run them over. In fact, he had a whole variety of methods developed for killing cats, from cars to guns to poisons to drowning. I had heard from Bart that he even stabbed a cat with a knife once while he was petting it. I never knew why Bruce did this. After all, what had a cat ever done to him?

When Bruce returned from his "vacation" in Montana, he soon became unwelcome in the house. Bruce and Ma had many shouting matches during the first few brief times he tried to live at home and it never worked out. He had rejected Jesus. Although Mrs. Kozy knew he was saved, she told me that he was simply a bad influence on the others and could not stay around.

He was different from the other Kozy kids. Maybe it was because Bruce was the oldest. Or maybe it was just a toss of the genetic dice. For instance, he was the only kid in the family who ever played in sports. As I said before, he was always the first in everything. The thing was, he was very mean about it. In fact, I hardly even remember any good conversations when talking with Bruce. He already had a go-kart when I was riding a tricycle. He always just called me "Fatso."

I recall there was one unusual conversation (actually more of a lecture from him). After he had his last fight with his mother and moved out for the final time, he drove past my house in a British sports car and started telling me all about how he converted this old MG into practically a dragster by putting in some sort of Buick engine and transmission. He had a girlfriend. He was living in a shed behind his girlfriend's house.

Later, I heard he had been accepted to go to Carnegie Mellon University on a full scholarship. I was wondering how he could pull that off—no one else I knew could qualify for a scholarship. Bart said that while Bruce was living in a shed at his girlfriend's house for a year, he declared himself financially independent from his family. He had no income and filed his own income taxes to that effect. He got the scholarship because of his playing baseball, and because of financial need.

Something didn't fit though; he still had his criminal record. Bart explained that Bruce had filed a petition with the California State Court system to have his criminal juvenile record sealed. My goodness, I thought, that's pretty smart. He was going to major in physics—he wanted to be a scientist.

Football

As time went on, our gang joined forces with another gang for more elaborate crimes. The other gang consisted of three boys from the Pierce family—Scott, Peter and Craig. This was a family of four boys, like the Kozys. There was an older brother, Dave, who had moved out of the house and was living with the father. Dave was a full-time construction worker, not part of the gang.

The Pierces hated football. The N-word and the J word (Jock) would fly out of Scott's mouth constantly. His younger brother Peter, although more timid, had basically the same vocabulary and opinions. Craig was very quiet.

One of our first adventures with the Pierces was to shut down one of the high school football games. The Pierces, coming from a bit of a wealthier family, drove the family Cadillac on crimes. Mark Kozy had access to a welding torch from Uncle Mitch's speed shop and we carried it in the back of the Cadillac. The two gas bottles, one for oxygen and the other for acetylene, were stood straight up on the back floor. We had a supply of helium, too, in a separate tank that we had obtained from the Stanford hospital, and a number of large balloons. Mark filled the helium balloon, and then he filled the second balloon with a mixture of oxygen and acetylene. The two balloons, tethered together, were released upwind of the football game. A two-foot fuse, obtained by mail order from a scientific supply house, was used for time delay.

The balloon set was released and Scott Pierce drove the family Cadillac away, followed by Mark, Bart, and Kurt in Mark's pickup followed Craig and Peter Pierce and myself in the Studebaker. Peter said, "You are not going to believe what's going to happen." We pulled over. The weather was a bit misty and you could hear the marching band play and see the lights of the football bleachers off a few blocks in the distance. The burning fuse in the night sky kept us alert as to exactly where the balloons were.

The sound of the explosion was incredible. There was a huge gasp from the crowd, then screams. People started leaving the game. We did not stay around to watch—we drove to the other side of the city to celebrate our success.

"How did that even work?" I wondered. At the Pierce house, we constructed a smaller test balloon, a demonstration, maybe 3 inches in diameter. The sound from the little balloon was ear shattering. It seemed like a stick of dynamite.

My mind was racing. There had to be some way of making use of this new discovery. Planning was called for. We went into the kitchen, but the planning was disrupted because Bart got into an argument with Peter about whether the turbocharger that we liberated from the Navy would fit on a Chevy engine. Bart said that any turbo would fit on any Chevy. Peter countered that you had to have the right manifold and modulator.

Then Mark spoke up, "Terry puts those blowers on small block Chevys all the time."

This issue was settled for now, because Mark had spoken.

Scott and Peter, the two older Pierce brothers, got into conversation, but it was more of a mock conversation, intended to draw the wrath of some others who were there.

Scott said, "Religion is just a bunch of bullshit. can you just believe how stupid all these religious people are with that God and Jesus stuff?"

"Singing songs about God in the heaven above," sneered Peter.

"Might as well sing a song about the tooth fairy!" shouted Scott, and laughed out loud.

Mark and Bart were silent for a moment. Then Bart started the counterattack. "If there's no God like you say, how did we all get here?"

"Evolution, idiot," shot back Scott.

"Shut up, that's bull," said Mark.

Bart said, "You cannot prove that humans came from monkeys. I mean have you ever been to a zoo and see one of those things? Our ancestors? Ridiculous. Evolution is stupid."

Mark had gathered his wits for a few moments. "You cannot explain why gravity is exactly the right amount of force to hold us to the planet without being so strong as to crush us all to dust."

Then Scott said, "What a bunch of crap, who the hell are you to make me explain that? You're out of your mind, Mark!"

Mark continued, "There is the matter of the heat of the planet. Why are we not burning up or freezing? Why is there just the right amount of oxygen in the air?" He leaned back into his chair. "There is only one explanation for all of that, and it is God."

"Satan!" said Scott in a loud voice. "Satan is the explanation." Peter and Scott just kept laughing in a loud, forced laughter. Mark was cold, icy, and stared straight ahead. Bart said nothing. I said, "There can be evolution and God, too." I must subconsciously have been trying to hold the crime family together. Now, I think I was foolish to say that. The fact was, we were also a debating society, and I should have just enjoyed it.

"Well, we might as well hit it," said Mark as time had just flown by and it was almost eleven by then.

Dave, the oldest of the Pierce brothers, had just pulled in the driveway in a white Chevy pickup. Mark's truck and my Studebaker were both blocked from leaving.

"How are you doing, Dave," said Craig, with great admiration.

Dave said, "Good as always. How are my little brothers? How are you Scotty?" he put his hand on Scott's shoulder in mock friendship.

Scott pulled away, "Don't touch me, you bastard."

"What's the matter, we're brothers, aren't we?" said Dave.

No answer was the answer. Dave had a case of beer and a quart of vodka with him.

"Fucking Alky!" said Scott. He stomped out of the kitchen.

It was up to Mark and me to talk Dave into moving his truck, so we could go on our way. The Studebaker started reliably. It had always been a friend to me. Mark still called it "the rust bucket." We drove home separately; Bart and Kurt were in the cab with Mark. The Chevy had about five times as much horsepower under the hood than I had in Stu. He came from behind me on Central Expressway and opened the throttle. By the time he passed me, he was going twice my speed. In a few seconds, all I saw were two taillights in the distance.

The Experiment

While Bruce was away at Carnegie Mellon University, Mark, who had no scholarship, worked part-time in the speed shop and started the first year of junior college. He took a couple of courses in astronomy and physics. It was interesting to me that Bruce and Mark never spoke with each other while in college. I certainly never spoke with Bruce at all, but Mark ended up being my companion. Sometimes, he would call me on the phone for no reason. Well, actually, there was a good reason: I was a great math student, the best student in honors calculus in high school. Mark loved astronomy, but sometimes had trouble in math, so I helped him.

My interest in motorcycles had moved to larger displacement racing—that is, to open class motocross bikes.

Bart took out a loan from his Uncle Milt to set up a small machine and welding shop in the garage and was making plenty of money while still in high school. My motorcycle was often at the shop. Younger brother Kurt was assisting Bart—and I don't know whether Bart was paying him or not. Bart also was into motocross, and he would win races. I just was lucky when I finished without killing myself (Being fat is quite a handicap in dirt bike racing).

Mark had become ever more consumed with physics questions in his second year of college, and, although he was a capable machinist, he kept away from the shop. At most, he would walk through the shop when Bart and I were working on a project. "What the hell is that shit" would be a typical comment. Bart and I thought we were doing some good projects. We didn't need that.

One time, Mark was carrying around a "Stellar Explosion" issue of *Scientific American* magazine. I got curious, so I asked him what it was about when a star blew up.

"Kaboom!" he said, and laughed.

"Mark, cut it out" I said, feeling hurt.

He said, "It's just another way God makes miracles." Then he switched the subject.

"Have you heard about the compaction of ammonia into crystals like they were doing at IBM?" I said I hadn't. I didn't read any newspapers. "They say that the energy released upon decompression is six times the energy in a stick of dynamite per unit mass." I still didn't know what he was talking about.

But I did have another idea. "Mark, that old refrigerator that your mom is storing in the garage. It's just been sitting there for years. I've got a great idea."

The idea was put to test in the parking lot of Ford Aerospace at about 3:00 AM the following Thursday. We drove the Studebaker over to just behind where the big antennas were and unloaded the old, unworking refrigerator to the asphalt. We had the oxygen and acetylene unit, but did not need a helium tank this time. Instead of balloons, we just filled up an extra large leaf bag with the gas, placed it inside the refrigerator, and closed the door with a nice long fuse leading out. A large strap was wrapped around the refrigerator for some extra compression. Bart lit the fuse and Mark kept himself a good distance away.

Bart, Craig and I got back in the Studebaker and drove away to a safe distance. It was a very quiet and starry night and those antennas were impressive to view—they must have been 40 feet in diameter. In the background was the light from Moffett Field. The only sound was the sound of a few cars passing on the freeway in the distance.

The detonation was breathtaking—it truly scared us. I have never in my life seen Bart in such a panic either before or after. Not only was there an explosion at the location of the experiment—there was a second, small explosion in the air above the first one. It was not clear exactly what the second explosion was, so we just stood there staring, while all the lights in the Mountain View and Palo Alto area went out. It was complete blackness. The debris from the refrigerator had landed in the high-tension wires off to the side of the parking lot. The parking lot was at least the size of a football field, and we had not considered that this could have happened. But it did. A few more moments of shock passed. We heard alarms going off, and then sirens. It was time to go home.

At the Kozy house, I remember talking for a minute about the blast, and then each of us going our own separate ways very quickly. I had to get home—we had school the next day.

Getting to bed at 5:00 AM just to wake up at seven to go to school can make quite a rough day. We, the junior members of the gang, usually met in front of the metal shop at ten o'clock, but this day was different. It seemed half the high school knew what we had done. Jim Waterford passed me in the hallway and simply said, "Boom!" School was not shut down; the electrical service had been reconnected by 6 AM. When I saw Craig, there was first a nervous chuckle with shifting eyes. Then Bart walked up and the laughter grew and grew. We all laughed until we were sore and could not laugh anymore.

Chemistry class in the afternoon had a different feel to it than before. I had only gotten an hour or two of sleep that night, but never felt so wide awake. Mr. Jamison spoke about the oxidizing power of some sort of chemical (potassium permanganate), and I listened to his every word. There were acids and bases, oxidants and reductants, and it was all very interesting. Chemistry had a very clear purpose, and that was to make explosives.

There was a quiz in chemistry class that Friday. I remember finishing early and just sitting there, thinking maybe we could go to Stanford pool later that night. Or maybe we could plan some motocross riding that weekend. Mr. Jamison wanted to know why I was not working on the exam.

"I'm done," I said, handing him the completed quiz.

He took the paper and looked at it for a moment and did a double take.

"You wrote the solution," he said in a whisper. I asked him what that meant. He whispered, "See me after class."

I was wondering what he wanted. Maybe I was in trouble. Maybe he knew I was part of the gang that had knocked out the power earlier in the week.

He said, "Stan you are amazing—I have been giving this same quiz for over twenty years and no one has ever seen a perfect score, let alone after finishing early. I just wanted to congratulate you."

I walked out thinking I was really something special. Not only I had been the original planner of the experiment at Ford Aerospace, but I had gotten a perfect score on the chemistry quiz, and the teacher liked me.

Christina

By that time, my expenses had grown to the point where I had to have an after school job. The paper route was just not enough. Besides, having a job after school was simply what a guy did. Jim Mason had a job and he had a really cute girlfriend. So it made sense. The job was at a pizza parlor, a great way to eat all the pizza you want, which I did. The boss, Mr. McCloskey must've weighed 400 pounds and he just loved cheese pizza. It was kind of strange, but he never had anything but cheese pizza. He ate a lot of it, and he encouraged me to do the same.

My weight grew to about 250 pounds, almost all of it fat. It was impossible to make any sports teams. I was in intramural wrestling, and there were only two other kids in the school who were in my weight class (the unlimited weight class). They were the first- and second-string middle linebackers on the football team. With Quinn, the second string linebacker, I could last maybe thirty seconds before I was pinned. With Ring, the first-stringer, I could count on him pinning me in five seconds or less.

There was one other "sport" I was involved in. But what the hell kind of sport is it where no one shows up to watch, and no one other than other players care whether you win or lose? The "sport" is chess. Not only is it a parody of sport, but, in some ways, chess actually trains a person to approach life in completely the wrong way. If the reader of this story has played chess, he will know why. No game that mimics life would ever be played having zero undisclosed information. In other words, what's wrong with chess is that all the pieces are on the board at all times for both sides to see. There are no hidden pieces. It is even more silly a game than playing poker with all the cards face-up. You play the game literally believing the universe revolves around outsmarting the other person! Good luck with that in the real world. The better of a chess player you become, the worse off at real life you will be.

The prep room in the pizza parlor is a great place for chessie or any other type of introvert to have an afterschool job. The instructions are clear. Show up. Follow directions. Fill the plastic containers with chopped veggies or whatever. It is really hard to blow it.

In order to save space, the pizza shop had two prep cooks working at once. There was a boy, namely Stan, cutting on a cutting board facing a girl, Christina. Everything happens in divine order, right? For instance, a star blows up and an entire solar system is erased from the galaxy. A boy meets a girl and he thinks she is pretty. A girl meets a boy and thinks he is ugly because he is fat. It was all in perfect divine (or maybe Darwinian) order.

It was springtime; as I had mentioned, I had already successfully planned and executed a large test at Ford Aerospace and gotten a 100 percent score on the chemistry quiz. It was all going just right. Now, Christina was going to be my girlfriend.

"Go out with me," I said.

She stopped. She was red in the face. The sixteen-year-old blonde girl had such fair skin that she could not hide a blush no matter how hard she tried. She was bright red and said simply "No."

I was quick. "How about tonight?"—as if her answer had been otherwise. She provided no further answers and returned to cutting the vegetables. The entire universe was collapsing down to a single point. The point was Christina. I was speechless. I tried to cut more veggies (or whether it was salami I don't quite remember). Whatever had been important to me up to that moment no longer counted. The only thing I cared about was what she thought of me.

She obviously needed more convincing, or some talking to. I may have stopped breathing completely at that point. My actions on that day were unlike any before. She only lived a few blocks from the pizza parlor. She had always walked herself home after work, yet it suddenly became necessary that she have a ride home. The question "Can I give you a ride home" was all I could think of. I was playing chess, behind a few pieces, about to be checkmated.

The answer was already no while I was still asking the question. She walked down the roadway towards home. The old Studebaker was in the parking lot. It was my pride and joy still. As always, it started reliably and soon, I was next to her on the roadway.

"Are you sure?" I asked.

She stomped her foot down in disapproval and said, "If you must," opening the door and getting in.

We drove in silence to her house while I tried to not to grind the gears too much while shifting. We reached her house on Homer Street. I told her how much I really liked working with her and I how I'd like very much to see her the next day at work. I cannot remember if she said anything at all. The Studebaker made it down Central Expressway that Friday evening after dropping her off at the house. The thing did smell funny. Not only was I the fattest kid around, but also my Studebaker smelled bad.

Central Expressway led all the way to the Kozy house after maybe a turn or two. Bart was working in a shop on his lathe and it was one heck of an important project he was working on. I put a face mask on and planted my face in close to the turning metal as I could so that the metal bits were hitting me right between my eyes on the mask.

When I got home that night, I told myself that certainly, I shall feel better by Monday. Saturday morning arrived earlier than usual; I finished my paper delivery, which by now took place via Studebaker truck. I came home, changed back into my pajamas, and climbed back into bed. I had never done that before. Then I closed my eyes. That made it all go away just for a moment.

So I pulled the covers over my head to try to make it all go away—really go away. I saw my feet down there. I started thinking about explosives. Motorcycles. Sports. Maybe that would work, I thought. It worked for about thirty seconds. Since I'm lying here, this will be a good time to catch up on my Bible reading. Let's start with Genesis. As soon as Adam and Eve were in the garden, I was with Christina in my own dream world. Who cares about Genesis? I tossed the book into a corner.

How do I approach Christina next? What strategy do I use? I spent the weekend staring at the ceiling, interrupted only by the need to deliver newspaper Sunday morning. I sought the solution to my problem. She had said no, she did not want to go out with me. Yet I must convince her otherwise. How do I see her, to talk to her again, considering that I cannot go to work again until Wednesday?

Monday at school first started out like any day. Then new things became apparent. The girls all look the other way when I came by. I also noticed suddenly that Pete, Byron, and Reese all had girlfriends—and I didn't. Walking from one class to the next, I heard a new word come from my lips—"depression"—a word I had never spoken before.

Wednesday came around and I wanted more hours working at the pizza parlor. McCloskey was negative and had even taken my name off the work schedule. He told me it just wasn't working out in the prep room.

He said, "We thought it would be good to have a boy and a girl working together on prep, but it looks like we were wrong."

I was eighteen years old, and had in fact been fired for sexual harassment, probably putting me on the leading edge of a new movement. I guess I could have been proud of being avant-garde. From my young eyes, however, things look nothing but bleak. All I thought about every waking moment was Christina.

I went to the prep room. There, Christina was talking to Jimmy, the owner's son, an older guy of probably twenty. She was giggling. He was teasing her about not having a boyfriend and how she was probably ready to have one. She saw me, and she stopped laughing immediately and looked down.

"Hi, Jimmy," I said to him. There was an awkward pause.

"Hi, Christina" I said to her.

She was all of a sudden busy chopping vegetables, but she was able to say, "Hello, Stan."

For a moment, I thought I had something to say to her, but Jimmy was standing there. I thought twice about it and started an awkward conversation with Jimmy, one like I had never had before. It was as if I had to just say something, anything at all, to make my lips move and break the silence. He somehow seemed to sense the predicament that I was in. For some reason, he told me that he thought I was a good guy. Nothing more. I left the prep room and went home, never feeling so alone in my life.

I dragged my way through the end of the last semester of high school. Sitting in chemistry class was a good-looking girl, and I had hardly noticed her until now. My God, I thought, I had even sat next to her in ninth grade math class. She used to tease me about being smart. Now she was the most popular girl in the school and was the quarterback's girlfriend. He would come by at the end of class—and kiss her right on the lips in front of everyone at the end of class every day. The yearbook was being written. Did I have anything to add? Blank. Crumble up. Toss away.

Graduation

After a week or two had gone by, I noticed I hadn't eaten at all. This was probably a good thing, I thought. I turned eighteen, so it was time for my driver's license picture and weight measurement. I had lost 25 pounds. "Now, the girls will really think I'm cool," I thought. Although some of the girls have boyfriends, Kathy, a girl that I had known from elementary school, didn't. Maybe she will go out with me, I thought. Her last name was unusual—there was only one in the phone book.

"I want to talk to Kathy."

"She's not here, can I take a message?"

"Have her call me."

"Who is this?"

"Stan."

"Stan who?"

"Hall."

"Who are you, a friend?"

"Yes, a friend." I told myself this is a good start. Then I realized I had not taken a breath in probably four minutes, and was dizzy, almost fainting. I left my phone number and she never called.

The chemistry class indeed influenced me, so I gave my final English report on four chemistry books I had read recently, all dealing with explosives. I stood in front of the class and described how to make gunpowder, nitroglycerin, and then the atom bomb, followed by the hydrogen bomb. It all seemed pretty simple and I was rewarded with an A for the work.

My parents already knew that it was the other direction for me. That is, while Stanford is on one side of Palo Alto, Fremont College (a community college) is on the other side. It all made perfect sense: no tuition. Cheaper. I figured it was the same thing as Stanford. Mark was there, we could work together and he got me a job in the speed shop after school. That paid well enough so I could quit my paper route.

By the time school started in the fall, I had plenty of money from bolting things together, and had lost about 60 pounds. It seemed like great time to start bike riding, exercise, and more. Mark was working on racing engines. He also was building physics experiments, even through summer. He showed me a detector device for solar radiation. He was full of long explanations about the flux of radiation at the Earth's surface versus time of day. In turn, as the local expert in mathematics, I would lecture him about how to perform calculus operations properly as needed. He would lecture me on just about everything else. The only time I would really listen was when he spoke about racing engines, for by that time; he had started working the races at Fremont and was constructing his own dragster.

Class work proved to be too easy—right away I got another job as a pizza cook. In my world at that time, this was a big advance over merely doing a prep job. I had made the big time. I had the assembly work in the mornings, combined with pizza cooking at night, giving me enough money to buy a new Dodge truck.

The second semester of college started with me running into Christina, the blonde from the prep room, at 8:00 AM the first day. Although she had not graduated from high school yet, she had started college classes. I spoke with her like she was my dear long-lost friend. Luckily for her, she was with Donna, her friend, to talk to as if I was not there. The fact that I'd never stop thinking about her for one minute since the day I was fired from the prep job was dearly private to me.

I thought certainly she would notice that all the pounds were missing. I was 60 pounds lighter, wearing the same clothes from before. Certainly, she would notice. She apparently did not. But there was one important piece of information she gave me. Her classes started at 9:00 AM and ended at 10. I could just always be there right in time when she was getting out of class.

The next day, as planned, I was right there. Donna saw me first before Christina, and turned her body so as to run interference. The two of them slinked by without acknowledgment and I was speechless.

There was a walking path from the classroom heading across the parking lot to Donna's car. The next day at 10:00 AM, the new Dodge truck, with me in it, was right on the pathway. Certainly things were different now. After all, that old Studebaker was long gone. The interaction was mostly the same: Donna ran interference so Christina would not have to look in my direction. They sped up there walking and intensified their talking. My window was rolled down and my arm was extended, waving as they walked past, to no avail.

An Accident

The beautiful Dodge pickup took me home in the middle of the day and no one was there. Mom was on an errand. The house was different. It was way too small. It was mom's house. Not mine. The assembly work at the speed shop was all caught up, so I had nowhere to go. There were no distractions available. The pizza job did not start till 5:00 PM. There was a new level of boredom and anxiety and I had to do something.

Bart was at home running his machine shop, so that was the place to go. I found him and Kurt working a large set of aluminum parts. Frantically behind schedule, Bart assured me it was nothing personal but he just had to keep working.

"We have both been up since 2:30 this morning. That's how to get things done without distractions. In this place, the phone starts ringing at 6:00 AM."

Sitting in a chair watching them cutting parts was mildly entertaining for a while, but I decided to go home and get my motocross bike—the one I still had from high school. I was so restless I decided it was a good idea to bring it over to the Kozy house for some modifications. At 3:00 PM, I was cutting the frame with a cutting torch, trying to improve the shock angle. After all, this was the latest trend in dirt bikes, a cantilever type suspension with much longer travel (I'd read about it in a motorcycle magazine.). It was a fairly simple operation, requiring cutting off some brackets, moving them lower in the chassis, and then rewelding. Bart could help me with the operation later, which would require a special welding technique.

I got comfortable sitting cross-legged on the floor and I began the cut. There was a noise in the corner of the shed where I was working—maybe a bird had flown in. I couldn't see too well with the welding goggles on. I took them off and spun around to my right. My left shoulder caught the handlebar and kicked the front wheel to the side so the bike fell towards me. Since I was afraid that the hot metal I'd just been cutting would hit me in the face, I shielded my face from the bike with my right hand—the hand holding the torch. The bike tipped towards me and continued to move, forcing my right hand into my face.

The important thing was to turn the cutting torch off. Spinning around on all fours then, it was possible to lift myself up and away from the bike. I remember clearly looking down on the torch lying on the floor thinking that I had better make sure the gas was turned off because we didn't want a gas leak—that could be dangerous.

I walked into the machine shop to tell Bart and Kurt about how the motorcycle had fallen over. When Kurt saw me, he started screaming. The machines were shut off and cuss words were soon flying out of both of their mouths.

"You are going to the hospital right now!" Bart said.

He got his car keys out. While driving in the car to the hospital, I began to scream uncontrollably. I remember seeing Kurt as he looked back at me, completely terrified.

The hospital was only a mile away, on El Monte Road. Bart just pulled his station wagon right into the emergency entrance and I was inside almost right away. I could still see out of one eye and caught glimpses of hospital staff running back and forth obtaining equipment. I was told to lie down on a cart in a room, separated from the other patients only by a curtain. Apparently, I was not too cooperative, for after a few minutes, two of the hospital staff arrived and put my arms in restraints. I was strapped down to the cart.

My mother and father were at the emergency room almost immediately, even before I was admitted to the hospital.

I was wheeled into surgery to stabilize the situation. I think they were cutting off chunks of flesh that could not be salvaged. The pain was indescribable. The process seemed to go on forever, but it was probably only after a few hours I was wheeled into a regular hospital room. I was set up with some IV bags and some machines looking at me. A nurse visited me every half hour. The pain relievers they were giving me only helped up to a point. I often wanted to scream but didn't.

Over the next few days, my parents regularly visited together, although they remained divorced. They both told me over and over again how sorry they were. I did not know what either one of them meant by that, and I did not ask. I noticed that my dad did not seem drunk.

Bart and Kurt visited often, and told me about how the shop was doing. It was a high point for me. Even after the extreme pain was gone, I was bored to tears.

On their last visit, Kurt said that his mom Midge thought that demons had made the accident happen, to try to get me to renounce God. I just listened and said nothing. But when they were gone, the words came to my mouth: "Fuck you God." Saying it once made me feel good. So I said it over and over and over to myself.

After two weeks, I was permitted to go home. Bandages covered the left side of my face and for some reason, I had two black eyes. I guess I was lucky, I thought. The torch could have burned out one eye or both of them.

Mark came by. "I need help with some math." He sat down smiling. "Just kidding." He had a large grocery sack with him, set it on the table, and pulled out four quarts of beer.

"Two for you, two for me." Silence was the only response from me. He then continued,

"Stan, you take four beers and divide by two people and you get two beers per person." Oh, he was trying to be funny, I thought. Beer math.

"The doctors told me I cannot drink—not with the pills they have given me."

"Oh, well," he said in altogether good spirits, "that means four for me," twisting off the top of one of them. I could barely move my head in either direction; the burn extended from just below my eye down to the tip of my shoulder.

"I have figured out a new way to kill a cat," he said. "Phosphorus. Works great. Will burn all the way through the cat from one side to the other."

He was still showing his lighter side, smiling and trying to cheer me up.

"I have been reading *Soldier of Fortune* magazine. These guys who served in Persian Gulf, they are telling me some of the greatest stuff about how they used to kill ragheads." He said, "A 20 mm round right through the stomach cleans out everything, leaving only the head, shoulders, and two arms!"

I felt like either drinking or throwing up listening to this.

"Get the hell out of here!" I yelled.

My mom came in from the other room and said to Mark that maybe it's time to leave. He did. He got on his motorcycle with the 4 quarts of beer (One of them half finished), tucked them in the saddlebags, and motored off.

My mother assured me that the pizza parlor certainly wanted me back as fast as I could come back. "They're keeping the job for you, and they think you are a really a good pizza cook," she said. We ordered pizza to go from the place, but I wasn't too hungry. Besides, I could not chew too well. The burn constituted another method for losing weight, a very convenient one.

I missed the rest of the semester of school, of course. Until the beginning of the next semester, I was able to assemble wheels at the speed shop. The business was slow, however, so there was only 10 hours of work a week. My face still hurt like hell, and skin graft operations were planned.

One day while I sat in the living room and watched my mom prepare to go shopping, I had the most startling thought. No one had visited me except Bart, Mark, and Kurt. What about Bruce? I thought we were friends. He was finished with his bachelor's degree from Carnegie Mellon, but he was back here at Stanford now. Mark gave me his phone number so I called. Amazingly, he answered on the first ring.

"Who is it?"

"Hi Bruce, it's Stan Hall."

"Oh, yeah," was the reply, followed by silence.

I told him about the accident I had, the recovery, and that I was at home sitting there doing nothing. He replied with a long silence, followed by "I need to go. I'm busy," and he hung up.

After I hung up the phone on my end, I had a feeling of pure sickness, pure loneliness, and pure boredom. I felt muscles in my chest started to twitch. These were muscles just below the rib cage, near the solar plexus. I could not stop shaking.

I had to do something, I figured, so I walked over to the window and looked out on the street. I thought about Christina and how she would be horrified to see me with this bandage on my face. How she heard the story of how it happened she would really care. But then when I looked at the lawn in our front yard, which was freshly mowed, I knew she did not care. I went to the refrigerator, where my mom kept some chilled wine. I poured myself a tall glass, and it made the shaking stop.

Junior College

After a while, the pizza parlor took me back, and my coworkers cheerfully welcomed me. I found myself looking forward to the end of the shift, when all the cleanup work was done and the beer tappers would open up. I still looked awful with the scar from the burn. The skin graft operations were of little help. But I did notice that no one made fun of me for being fat anymore.

We poured ourselves a few—Ross, Rudy, and I. They had girlfriends and I had just a story. The manager, Steve, brought down a bottle of Jack Daniels and gave us each a shot glass. Mine was gone and I had another. After two, the loneliness was gone. These people were my friends.

Rudy and I recalled how we had raced through the Christmas tree farm near Moffett Field when we were fifteen to get away from the cops. The hundreds of pounds of scrap metal we have stolen from various businesses in the industrial park. The successful heist of the Studebaker truck, and the attempt to get the Caterpillar diesel. "Oh, I was so stupid. I thought I could jack up a tractor with a floor jack in the mud!" he yelled, and we all laughed. Even Ross and Steve, who were not there when it happened, laughed out loud.

I hadn't felt so good in years. It was like old times. "Hey, where are Bart and Kurt now?" I thought while we exchanged more stories. I got another drink. Rudy and Ross had only gotten halfway through their first one. My third one went down.

"Where is Bart?"

"I have his phone number."

Bart answered the phone very soberly.

"We are so far behind on work I cannot even tell you. I have been working since 2:30 AM and now it is almost midnight. We will probably go to sleep here in another hour and start again at 6:00 AM."

I had what I thought was a great thing to say. "Bart, come over, let's drink!" Bart had come to be about as much fun, they say, as the proverbial turd in the punch bowl. The conversation was short and we went back to what we thought we were doing, and so did Bart. I have no memory of the remainder of the evening.

The next day, Rudy and Ross were telling stories about what we had done. I listened to the stories and the two of them laughing about them, but I could only faintly remember the things they were talking about. I was feeling quite ill.

A Real School

After the next semester concluded, it was time to go to a real school to complete the upper division. Bruce was at Stanford. Mark, not having a scholarship, had decided to attend Berkeley to save money on tuition. He also had an array of projects nearby in Palo Alto and Sunnyvale, including his drag racer, solar detector, and cellular array, and a large plastic remote control airplane. The girlfriend had disappeared. The projects went into storage.

Visiting Mark in Berkeley was an unusual experience that I was not prepared for. He was upset about the course load, whereas the course load had never seemed to matter at Fremont College. After I'd been there for only half an hour, he asked me to leave, because he had to study for an exam. It was a little disappointing, seeing that it had taken over an hour just to drive there.

A few pieces of furniture were all that went to Berkeley with me. The distraction of moving kept my mind away from the despair, boredom, and loneliness. I still wondered about Christina. Would I ever see her again? Oh, my God, I'm in love. That's what the problem is. Love was supposed to be a wonderful thing, but it had ended up being the worst experience of my life.

The apartment I would be living in was several miles to the west of campus, in the flatlands near San Pablo Avenue. San Pablo was a four-lane street with traffic lights at every intersection. It did not feel like a college town there. The flats and apartment houses along this street tended to be very run-down. There were idle men wandering everywhere, some of whom were pushing shopping carts full of junk. Beggars were abundant. In the evening, there would be black women walking the sidewalks in metallic clothes, wearing blonde wigs. Many of the cars on the street appeared to be pimpmobiles. The rents were much more reasonable in this part of town than nearer to campus.

My mother helped me move the furniture into the apartment. We put the knives, forks, and spoons in the kitchen drawers and the bed in the corner of the bedroom. It was about three in the afternoon when she left, so I looked over the course work that I've signed up for. They were Chemical Engineering 142 and 160, plus Chemistry 127A and 12B. The one-paragraph descriptions were a bore. In fact, I was so bored that I went to a nearby liquor store and bought a motorcycle magazine to read. However, I could not even read one page.

Looking at the wall, there was a horrible feeling. The feeling had been there before, but this time it was very strong. There was nothing I could do to stop the muscles in my solar plexus from twitching, so I lay down in bed and pulled the covers over my head. The sun had not even gone down yet, but all I wanted to do was sleep.

The alarm clock rang at 6:00 AM. It's always hard to get out of bed, but I stood up and walked across the room. The moment my hand touched the alarm clock, I said to myself, "Today is the day I kill myself." Then a second thought came immediately—that I couldn't do that to my mom. So I then decided I had better not go near the knives in the kitchen.

It was safer in the bedroom, away from the kitchen, until it was time for class to start and then I would leave. It was raining and was about 2 miles on bike to class. I didn't even know where Gilman Hall was, so I had better leave early, I thought. Any further decision about killing myself would be made after class.

The first day came and went. It was clearly a case of things completely out of control. Halfway through the morning, I discovered my fly was unzipped. I had bought no pens or pencils, and had to keep borrowing to take notes. And, by the way, I had to take notes like my life depended on it. But I wanted to end my life.

Coming back to the apartment the first day, I called Mark, who was across town starting a master's program in the graduate school in physics

"What is it you say your major was?" he said.

"Chemical engineering" was my reply.

"What is that?" he said mockingly. Then he said, "Not only is that a stupid subject, but it is the hardest major there is. In chemical engineering, you will spend your entire life in the library."

I replied only with silence.

He continued, "Do you know when a chemical engineering student finally gets out? That's when he jumps off the top of the Tower in the middle of campus."

Since this conversation was just on the phone, I could not tell if he was joking or not. Mark had changed a lot and his voice was no longer comforting to me. I said goodbye and hung up.

I spoke with my faculty advisor, Professor Newman, who told me that this year was worse than normal, the department had accepted an unusually large number of students, and so had to do more "weeding" than normal. This process, "weeding," eliminated the students who were not serious. I had somehow signed up for all the "weeder" classes all at once.

By Friday of the first week, there was no question that I would certainly flunk. There was a chemical engineering problem that I simply did not understand, and the professor blew off my question in class. Worst of all, chemistry 127A had begun an adventure into a nightmare called "quantum operator theory." This was some sort of mumbo-jumbo designed to intimidate the students and make them hate the subject so that they would change majors.

I felt my solar plexus fluttering again as the first set of midterm exams approached. I had done all the studying possible. My brain would not quiet down. I could barely sleep. There was no chance of knowing what the hell a quantum spin flip operator was, so I just clenched my teeth together. By three in the afternoon, with classes over, the thoughts of suicide would arrive. Oh, yes, what about fun activities on campus? As the ugliest person in Berkeley, I just hid. How about a trip to a bookstore? Someone might see my ugly face. Better not go.

After a while, the day came when the first stop after classes was a liquor store. The bottle was usually of cheap vodka, and the mixer should have been orange juice. That was what Dave Pierce used to drink. Real orange juice was kind of expensive and I was on a tight budget, so orange Tang would have to do. The first one tasted awful. But if I just got it down quickly, it wasn't too bad, and I stopped thinking about killing myself.

Chemical Activity

After a drink, it was clear what my real problem was. There is a certain, fixed relationship between chemical potential and chemical activity and I didn't know what that was. The exam was to be at ten o'clock on Friday, and I hoped he wouldn't ask that question.

I got drunker and started thinking "How could anyone be as smart as Professor Prauznitz? He runs around preaching chemical engineering like it is some great religion. But it is a horrible subject. The teaching assistant is Wally White. I wonder what he is doing for his research? I could see him at four after organic lab. Then I wouldn't have to get bored.

When I got bored, all I thought about was nuclear war.

"These people, the students, Professor Prauznitz, Wally White, they don't know what's going to happen and how bad it will be. They just don't get it. They go about their business and think there is a future. There will be an end to all this probably in a year or two, and there may not even be any warning. The TV news may say that there is a nuclear attack, or it may not. What will my mom think if they announce a nuclear attack?"

The first midterm arrived on a Friday and the alarm went off at 6:00 AM, I switched it off, and went back to bed trying not to think about killing myself. There was a slight hangover as usual. I pulled the covers over my face and I could feel the scar on my left cheek rubbing on my wrist. It reminded me of how ugly I am.

Arriving at class, I told my classmate sitting next to me, in a panicked voice "I'm going to flunk."

He replied in an angry voice, "Shut up."

The test consisted of three questions, which I answered immediately and turned the paper in. The professor looked at the sheet and said nothing. The score turned out to be a perfect score. The rest of the midterms went basically the same way.

Wally took me under his wing as an undergraduate research assistant and I spent many hours working with him. His research advisor was Prauznitz. This activity captured the remainder of my college years. I spoke with very few people otherwise and I did not really do anything but work. I drank the amount that I had to drink. I did not kill myself. I stayed inside as much as possible.

Graduation placed me at the top of the class, in a tie for top honors with two other gentlemen, both of whom went on to do PhD work at other schools. Very few Berkeley chemical engineering undergrads were asked to stay on as graduate students. But I developed an acquaintanceship with Prof. Prauznitz, who was interested in urban aerosol plumes (a fancy name for smoke) and debris transport (a fancy name for ash particles) after urban events (a fancy name for a fire). Yes, I know, it was not the sexiest field of study, but I liked Prauznitz. In fact, we got to be first name basis after a while.

One stipulation of my being accepted for the graduate program at Berkeley was that it would be for a master's degree only. If I wanted to try a doctoral program, it would have to be at another school.

Glenn Prauznitz had grown up in Germany and learned English as a second language, entering MIT at age sixteen. He graduated with his BS degree at the age of nineteen, entered Princeton immediately thereupon, completing his PhD at age twenty-four. Then he did postdoctoral study at Berkeley. He became a full professor at the age of thirty-one. He kept his private life to himself and we talked shop. Oh, boy, did we talk shop. Main subjects were the governing equations for fluid mechanics, conservation laws, and methods of convergence. He did have a home life, a woman he had married shortly after getting tenure, and they had two children.

My thesis simply was to be the extension of my undergraduate research on plume dispersion around urban objects. I thought it was quite a clever thesis—the theory predicted the size of particles downwind from a fire as a function of distance, wind speed, and altitude. The problem for my master's degree thesis was to do the experiment, to prove the theoretical prediction to be right.

The degree required me to take a few more classes and also teach a few. The project itself required the use of laser interferometers, which the UCB chemical engineering department had some of the finest in the world. A testing area on the other side of the football stadium was to be used. Small fires would be started (inside small, ventilated brick ovens) under different wind conditions, with differing fuels and the laser would be used to scatter light from the smoke particles travelling in the air downwind, to examine their density, size, and shape.

Having set the proper paperwork in place to check out the lasers, it was time to get them for "dry runs." This is a test without igniting any fires. We would then measure ambient conditions throughout a given set of days to determine background particle concentrations.

Karen

I had a good reputation at Berkeley. My weight was under control. I must've had a smile on my face when approaching the laser equipment counter in Gilman Hall that day. The blonde behind the counter told me to fill out a form and sign it. I did my usual thing, signing it while looking down, not making any eye contact. I looked at her from my right side, and I rotated my body so as to point the burn scar the other way.

"You are in theoretical chemistry 227B," she said.

I laughed—without knowing why.

She said, "I'm Karen, I sit right behind you."

Yes, I did recognize her. If I had been my usual self that day, I would've denied it. But instead, we talked all about Professor Hess and how weird his haircut is, what kind of accent does he have . . . Hungarian?

Then she asked me out.

Since I was sure this was a mistake, I said, "I'll think about it," and turned leftward and started to walk away. Then I stopped after a second and said "Yes." She thought it was hilarious and let out a huge laugh. While she was laughing, I noticed something unusual about her. She wore quite a bit of makeup for a graduate student. And she wore high heels and a pearl necklace, which was all a bit more than what I was used to seeing. I got her phone number and then went up to Sullivan field where the experiment was to take place.

It all went well that afternoon. Then I locked the interferometer in the Dodge pickup and went to the apartment to drink, planning on getting up early the next day to get an early start.

I got to my apartment and put on a hamburger and got halfway through my first scotch for the night when my phone rang (I had already sworn off vodka by graduate school; besides, by then I could afford to drink better.) I figured it must be Mom calling—she's the only one who ever called. It was Karen.

"Hi, Karen." I found myself smiling—and nervous.

"I thought I was going to call you."

She said that she was thinking about me all day long after she had seen me. She had gone to Prauznitz and gotten my phone number from him.

"How did your experiment go?" she wanted to know. It sounded kind of strange to me seeing that there were hundreds of experiments going on at Berkeley at any given day; then why did she want to know about mine.

"Let's see each other tonight," she said in a syrupy voice.

I had already started cooking my hamburger for the evening, so it would be wasted now and it was a shame to throw away food. So I started thinking about how to save it.

I guess I can just freeze it, I thought to myself.

She said, "Jaspers at nine o'clock?"

"Okay," was all I said. "See you there."

There was no thought of picking her up or driving or whatever. I still rode my bike most places and this night would be no exception.

I wondered what she wanted; something didn't sound quite right. I had another scotch. "She is simply neurotic," I said to myself. It was good that I had diagnosed her. It made me feel smart.

Karen was thirty years old, eight years older than me, and blonde. She was a very smart MIT undergrad with a 4.0 grade point average. Math was the easiest thing in the world to her. She wanted to tell me all about her first two husbands. Neither of them was any good at all—they were pathological liars and cheaters. She affirmed, however, that *she* was amazing. She was the state swim champion in high school. She was a great singer in the girl's choir, and had actually sung on TV when she was in high school. There was probably nothing that this blonde could not do well.

One couldn't help but notice the age difference. We were both in the first year of graduate school, but she was eight years older than me. What had happened in those nine years? She had first tried to be a housewife. After her first divorce, she became a massage therapist. When she got married again, she tried to be a housewife again. Now, she was starting school again after her second marriage had ended in divorce.

Did I mention she was blonde? There is one more thing I noticed. She was not a natural blonde. She colored her hair—I could see that her hair roots were black.

It was good that she showed up with a few drinks in her.

Shortly after we arrived at Jaspers and had a drink, she said "Are we going through the pretense of eating dinner, or are you coming over to my place now?" There was just a smile from me. My glass of scotch was empty, but I tried to pick it up anyway to take it to my mouth. My hand was shaking and the ice rattled in the glass, making me further embarrassed.

Until that point, I had stayed up all night on several occasions for a variety of reasons. The reasons included testing homemade explosive refrigerator devices, stealing Studebakers, driving to Los Angeles and back, etc. I had never stayed up all night for sex. It was unbelievable the first time—I think maybe that I passed out. She was screaming so loud I thought she was faking it or something. In fact, I didn't know what to think—I've seen porn movies but never really done it myself. It was as if I had been wandering in the desert for years without water, and then had found a fire hose to drink from.

The next day at the experiment on Sullivan field, there were two of us. The detector lasers were set in place and data collection began. Somehow, there was a reason to kiss, for just about any reason. First, it was because the experiment was set up. Then, we kissed because the data look good. Then we kissed because we found each other. Who would ever have guessed such a great thing could have happened?

She went home early and I had the afternoon class to teach, so we did not see each other for the rest of the day. I was out of scotch so I made the usual trip down to Ed's Liquors on Shattuck and bought another quart. Melissa, the office secretary, was getting a 24-pack of Budweiser, and said hello. She said that she and her husband were getting ready for the football game on the weekend, and they had friends coming over.

"Go Bears," she clenched her fist and pushed it into the air. It was kind of cute.

I got home and the phone was ringing and I answered it.

"You bastard," a woman's voice came out. It could only be Karen.

"What were you doing meeting Melissa at the liquor store?"

"I didn't do that."

"Yes, you sure did."

"How did you know that, were you spying on me?"

"I have my ways."

Then she said, "I want you to come over here right now and apologize."

So I did. At her place there was an open empty bottle of wine. She came running across the room and put her arms around me.

"Darling, never leave me again."

We went to the bedroom again. Afterwards, she got angry again when I tried to leave. Her face was completely red and she screamed "insincere" and "total asshole." I felt relief when the front door closed behind me.

Hiroshima

I said to myself, "There is somewhere I need to be right now and it's not here." So I went up to the Bancroft Library, where a special book featured in the front entrance, written by George Bunn—*Arms Control by Committee: Managing Negotiations with the Russians*—was the name. Looking back, it is sort of funny that I had never visited that particular library, which was the main library, in all the years in Berkeley, until that night.

They let me check out a copy of the feature book, and I read it. With my bottle of scotch in my apartment, I learned that night that it is possible to read an entire book on nuclear weapons, arms control, and treaty verification in one night. By around 6:00 AM, I knew that my life would never be the same.

The classes the next day were the same as always, but there was no Karen in chemistry 227B. Just a week before, she had been a total stranger. Now, I feared her. In fact I hated her. After class, I checked out two more books from the Bancroft Library. One was by Victor Utgoff, called *The Coming Crisis*, and the other was *Hiroshima*, by John Hershey. I had previously seen the photographs of Hiroshima after the bomb. There is a story in John Hershey's book about some survivors who were unlucky enough to be standing outdoors facing the bomb when it went off. Their faces burned completely, they wandered the streets afterward, unable to see anything because they had lost their eyes. They asked anyone they could for water or medicine. The author of the book described what it was like to look in these men's faces and listen to their pleas for help.

The weeks went by one after the next. Things moved ahead in the experiment and there was no Karen in class. I asked Prof. Hess about her and he said she had dropped the class for a medical reason. At that point, I went over to her apartment to check things out. Maybe she needed my help. I felt sorry for her. The ring of the doorbell did not cause any type of stir at all.

I turned around and walked back to Shattuck Ave. to get another bottle of scotch. Just as I stepped inside the liquor store, a car passed. It was her car, and she was driving. I got home to my apartment and I was planning another good read: *Ground Zero: Ultimate Disarmament*, by Amory Lovins. Karen's car pulled into the driveway and stopped. She got out of the car with a small revolver in her hand.

"Bastard, I hate you!" she screamed while pointing the thing at me with two hands.

To my horror, I saw her squeeze the trigger. The hammer went back then forward "snap." Gun empty.

"Ha, Ha, Ha!" she screamed hysterically.

Then she flung herself in my direction, pushed her face directly on mine, and put her tongue in my mouth. I was shaking badly and just want all this to go away, and I did what was probably just about the right thing to make that happen. I pushed her back, then calmly asked her to come inside the apartment. I found myself speaking words as if they were coming from a source outside of myself. "Karen, see that chair? Sit down in it." She did. Then I sat down in another chair facing her, looking at her right in the eyes.

I said, "I never want to see you again. Now, what do you have to say?" Strangely, she said nothing. Then I said, "I want you to leave and never come back." She went to the front door and she was gone.

Although I did not see her, I lived in the fear of her at all times for the remainder of graduate school. Eerily, it was like how I had never stopped thinking about Christina, years before, for days and days at a time. But when I thought about Karen it was mostly how I could avoid her.

It was time to get counseling because of the trouble Karen was causing me, so the student counseling center provided me this service readily. I had to explain to Dr. Judd exactly how I'd never had a girlfriend before Karen, and how she was a psycho and probably worse, an alcoholic. He recommended that I see him every week, but he had a strange unexpected piece of advice. He said that I should attend "Al-Anon" meetings at the Unitarian Church on Friday nights. Those meetings are like AA meetings, but for people who are instead tangled up with alcoholics.

The schedule of graduate school was altered a bit. After that, I divided my time between reading books on nuclear weapons policy, attending Al-Anon meetings, going back to thinking about suicide, drinking, not killing myself, working on the experiment, working on teaching, drinking, think about writing my thesis, drinking, teaching. Then I also took a math class or two.

The Prisoner's Dilemma

Near the close of the semester there was less time for reading books. However, the shortest of books just fell into my hands at a garage sale one Saturday morning: *Introduction to the theory of games*, by Ken Wilson. I was reflecting at that moment that what Karen and I had been doing was playing a game, a game where I did not really know the rules. But it usually turns out, I discovered by reading the book, that there are not fixed rules in most games. Many times, rules are nothing more than the rules of circumstance.

In the book, a *Prisoner's Dilemma* is described as two people locked in a prison cell. The warden says that both prisoners shall stay there for ten years unless one turns on the other and testifies. Then the one who testifies goes free and the other stays for the remainder of his life. However, if both testify against each other, the deal is off and they both stay for the rest of their lives.

There are variations to the game, but it is called a dilemma whenever *neither* one can safely rat the other one out and go free. If they are kept in separate cells, they will both assume that the other one is going to be a rat, so they each may become rats with that expectation in mind, and therefore both stay in prison for life. If only they had cooperated, they could have gotten away doing the ten-year thing. But "everyone knows" that won't happen, so it does not happen.

A much more interesting variation is when the dilemma is repeated as a series of small steps. For example, the rules could be such that one prisoner can add to the other prisoner's sentence by two years (and reduced their own sentence by two years) by telling about one crime. The prisoners can keep score, without communicating with one another, by inquiring at how much time remains on their own sentence. If his sentence has grown by two years, then a prisoner knows that the other one has ratted him out. Then he would retaliate to teach the other prisoner a lesson. The other prisoner, in turn, sees that his sentence has grown longer, so he retaliates. And so it goes until they both have over 100 years left on their sentences. Then the game is over and they both lose.

The rules change somewhat if the prisoners are allowed to talk to each other, and more complicated strategies will develop in negotiation and bargaining. The game is most interesting when it is a repeated prisoner's dilemma, with partial communication allowed. It is the set of possible strategies and how they play against each other, which make the game so important and interesting. Game theory, and the prisoner's dilemma in particular, also seems to describe many of the struggles ordinary people have in their ordinary lives. It is *not* just about prisoners in cells and a warden trying to play them off against each other.

The successful strategies developed for the real-world prisoner's dilemma called the Cold War are said to be the reason why we survived it. It obviously could have turned into a hot war with one side killing just about everyone on the other side. But there was the certainty that the other side would always retaliate.

Additionally, two sides stayed in communication by mutual agreement, and there were more steps leading to all-out war than just pushing a button. The more steps, the better for both sides, and the more communication the better for both sides. After years of living in the dilemma, the joint decision was made to keep the Cold War cold, because the consequences of only a modest retaliation, much smaller in scale than all-out attack, would still be so horrible for the recipient that the price—"victory"—would be far too high.

The "Cold War game" could have variations. One extreme example is to suppose it were discovered that none of our nuclear weapons would actually function when used. Just hypothetically, if the discovery were made and confirmed through secret testing, what exactly would be the consequence?

If there were no leaks of classified information, and our defense establishment took no action to signal there was a problem with our arsenal, absolutely nothing would change. Our opponent would still be deterred by the belief that our systems would work. In the case that they received intelligence information to the effect that our arsenal was inoperable, their decision-making would still tell them not to take advantage of the situation, because of at least two considerations. First, they would not be sure the source that had provided this "information" was a good one. Secondly, they would also be aware that the information that the United States had about its own arsenal could be due to an error that the US made in evaluation of its own arsenal. The cold war would not go hot.

In other words, the weapons themselves are not the instruments of the Cold War. The only "weapon" we really have is whatever belief the other side has about our intentions and capabilities.

The Oral Exam

December was when my oral exam would be and my job interview would be in January. The oral exam wraps up progress on the thesis research and the document itself should be fully written in draft form at that point. The thesis should be done before starting any interviews. After the oral exam, the final draft is written and the signatures obtained for completion. I ignored all these rules.

In early December, and vacation right ahead, the only way to get it done was to schedule it for Friday of finals week. Thursday, I had a final exam in a course in the math department, differential manifold theory. On Wednesday, I had to give an exam on quantum theory to the undergraduates. There, things went magnificently well and everyone passed the exam. During my own mathematics exam, my mind completely stopped working, and it could not produce answers to the questions. I got a D on the final, giving me a B for the semester.

The members of my thesis committee all showed up for the exam with smiles, and I spent the entire three hours telling stories to each other while they gave me some problems to work on the dryboard. While I was working problems, there was intermittent laughter. I was wondering why. Professor Hess was the chairman of the committee, and he said, "Okay, you pass," and there was more laughter. While they were shaking my hand, I wondered why they were laughing at me. I figure they must know that I hadn't really started writing my thesis yet. Maybe they had some other joke they were telling about me.

Actually writing the thing—the master's thesis— would take months. But because of my drinking and my need to go to Al-Anon meetings because of Karen, it was certain to take over a year. *Maybe I should quit Al-Anon*, I thought. I was interviewing for a job that was very important, and I couldn't start any new job until my thesis was complete. They must have known all this. That is why they were laughing.

It was 5:00 PM. I had my bike at school as usual and rode down to the Unitarian church, but I was so early that only Peggy, the chairperson, was there. "Peggy, I'm not feeling too well today," so I went to my apartment. The 12-ounce glass came out, this time just with scotch and no ice. And I got about halfway through it before stopping to take a breath. As it went through my system, the buzz from passing the oral exam became what was at the top of my thoughts. "Karen could not bring me down."

I took the bottle with me and drove my Dodge over to Mark's house in Fairfield, about 50 miles from Berkeley. Mark graduated with his MS two years before and had started working at Fairfield National Laboratory immediately thereafter. I did not really see him often. I pounded my fist on the door. Guess who was there? It was Dave Pierce, the oldest brother in the Pierce family, and he was making screwdrivers. A yell went out as the three of us were reunited.

"I passed my oral!" I yelled. "More shit! Let's drink to that!!" Dave said, drinking a 12-ounce glass of screwdriver.

Mark was drinking a quart of beer, and I poured myself a tumbler full of scotch, and began chugging it like beer. The good times were certainly here again. We turned up the radio on the oldies station and started swing dancing the best we could, the three of us drunken bachelors.

I woke up at my mom's house the next day. *Wow, what a night*, I thought, still having oldies music in my head. By about 7:00 AM, I realized that the party had been held at Fairfield but I'd ended up in Palo Alto. My Dodge was outside in the driveway. I went to the bathroom and threw up.

Mom was not awake so I just went on a walk, coming to the elementary school near her house. Flashes came into my mind of crossing the Dumbarton Bridge the night before with a hitchhiker in the car, the radio blasting. I remember hearing the hitchhiker telling me, "Man, you are crazy, crazy, drunk . . . . You just ran through two red lights. Let me out of here!"

The hitchhiker told me to stop and he got out and slammed the door. I didn't remember anything else about coming home. I started to cry. If I had hit anyone, they would be dead and I would be a murderer. The hitchhiker would be dead—that's two homicides. Oh, yeah, and I would be dead. That's three. Three strikes.

"God help me! I have to stop drinking. I must be an alcoholic . . . . No, no, I am not an alcoholic. Karen is really bad, she's probably an alcoholic. Dave is an alcoholic. My dad is an alcoholic, but not me."

I simply decided I was never ever going to drink again no matter what.

I walked back to my mom's house and she was awake. I went to the bathroom and threw up again. When I came out, my dad was there. He had spent the night, which was a bit of a shock. He said, "Rough night?" I ignored him, and thought, *What does he know anyway? He has never been through the stuff I've been through.*

Slowly, I made my way back to Berkeley, for there was plenty of work still to do. By noon on Sunday, the hangover had mostly gone away. I was ready to fight again. But first, I had to do something about this drinking.

The telephone book listed a phone number to call if you think you may have a drinking problem. The next AA meeting in Berkeley was at 2:00 PM at the Unitarian church, which was bicycle distance from my apartment.

A variety of strange people were there for the meeting. When it started, some poor soul stood in front of the group and asked if there were any newcomers. Of course, I would not say anything. Someone actually said yes, and they were required to stand up and say their name. Then they were required to say they were alcoholic, but I would not do that.

These people were quite slow while they were talking, and I was bored almost right away. Some guy talked about how he spent so many nights in jail. I thought how stupid. Why did you do those things that put you there in the first place? I waited for the hour to pass. Near the end, there was a gap of five minutes when no one spoke. It was so boring I just had to do something to keep everyone from falling asleep, so I stood up.

"My name is Stan, but I'm not an alcoholic." Silence (They had always said in unison, "Hi such-and-such," to every speaker before me.). "I'm here today because I just need to stop drinking long enough to finish my master's thesis."

They were still silent and I looked down and saw a woman's breast almost come out of its garment. Her orange straps were very loose and she looked like a junkie or something. *Why am I wasting my time on these people?* I thought. There was still silence in the room. They were staring at me.

"My thesis is about how the transport . . . how smoke particles . . .10 micron size . . . do assume size and velocity distributions . . . how smoke moves through the air and how big are the smoke particles."

I could not say what my thesis was about in any way these people would understand it. I was angry that they were wasting my time.

Then I noticed that the smell of the place was awful. It was as if 1000 people had already been in that room that morning puffing on cigarettes. I was furious that I was exposing myself to such dangerous carcinogens. I just sat down and said nothing. The meeting was over except for the circle of hands for the Lord's Prayer. It seemed totally bizarre to me that no one there had evolved beyond such a primitive thought system as that. I was out the door the instant the thing was over. I knew that the best thing to do was to ride my bike more often and faster. I need to get more exercise.

The job interview was on Monday. I had cut out all drinking now because what I was going to do at Fairfield was very important. The idea of experimental study of debris transport (smoke plumes) from fires is much the same as radioactive fallout movement from a nuclear weapon. This combination was just amazing to me. At last, I could really be someone important.

I would help the US move forward in the world with new arms control treaties. Nuclear weapons would be slowly reduced in size and number. I heard talk about complete elimination of weapons in our lifetime. I wanted to be part of it.

Fairfield Visit

Being inside the gates of Fairfield Lab seemed to me to be safe. Yes, it was good, in terms of economic safety—they paid well.

The first person I spoke with was Don Getz, the P-division leader, who asked why anyone would care about particles from smoke plumes. The "P" stands for physics, but he was effectively the head of the nuclear weapons design personnel. The right answer was that nuclear weapons turn cities into smoke, so we need to study smoke. The polite answer was that it was an interesting field of research with many applications such as forest fires, radioactive clouds from nuclear accidents, even biological weapon dispersion. He got the polite answer.

"That's not really research. Research is when you do something that is, well, interesting."

I looked at his Princeton diploma on the wall behind his desk.

"What did you do your thesis in?"

"Spin parity structure functions in heavy baryon decay."

I looked right at him and said, "That is what I call trivial."

He jumped an inch or two when I said that. *To hell with him*, I thought, *I'll never work in his division anyway*.

He said, "You and I are going to have an interesting time together," but I didn't know what he was talking about since the real point of this trip was to talk with the chemical physics division, not with P-division.

I just couldn't wait to get to the interview that counted. I sat there expressionless, across the desk from him, saying nothing. He cleared his throat.

"I guess we have nothing more to say then."

"I guess you're right," was my reply.

The next interview was in another building about a half a mile away in the newer section of the laboratory. Sean Kraus was the man I was seeing, and he seemed to be really nice. He knew Prauznitz and was aware of the importance of debris field measurements (a.k.a. smoke from fires). We talked about attending some of the same conferences over the years and he had just visited Berkeley last week.

He sat back in his chair and said, "You really do look like an ideal match for our division; when can you start?" We discussed how long it would take to write my thesis so I could start.

The End of Berkeley

My thesis took six months to complete. During the time I was completing my thesis, Bruce accepted a full-time position in P division at Fairfield. This is the very same division headed up by Don Getz, the guy I didn't like. It was not unexpected that Bruce would end up in P division, but it seemed unusual he would start at deputy division leader. I really did not know he had climbed the career ladder so much up to that point.

Mark worked in W-division, weapon engineering and testing, which had some cutbacks, and there was some concern about his position. The equipment he had been building was no longer in vogue. His knowledge of new nuclear instruments from his days in graduate school had gotten him "inside," but not exactly what he was hoping for.

My time at Berkeley was winding down, and the time spent writing consumed ever-longer portions of each day. Upon completion of a page, a small piece of my Berkeley experience would transform from the present to the past. The completion of a chapter would leave me with the feeling like it was time to leave town, like I was becoming more of a stranger to that town. I saw new students arriving and they looked so young, like they were just children.

The moment when I had a complete manuscript arrived at last on a Friday afternoon. All references and footnotes were in place and the proper cover sheet was attached (That cover sheet was available only at the

Graduate Office on campus and cost $150.00.) The document was about an inch thick and I ran three complete Xerox copies, one for each of the professors who would sign it. These copies were hand-delivered to the inboxes where they belonged. All of a sudden, there was nothing to do.

Something told me it was time to walk down to the main plaza on campus. The students madly shuffling from place to place looked so silly. There were the usual protesters against American imperialism, but looking very, very young this time.

I went back to the apartment and looked at the television for a while, probably about five minutes, but was bored. I lay down in bed and pulled the covers over my head. Although it was still daylight, this seemed the only appropriate thing to do. After a while, I noticed that it had become dark, so I got out of bed and undressed, climbed back into bed, and slept the rest of the night.

In the morning, I looked out the front window and the apartments around me seemed completely empty. There was a flat across the street, where sometimes a Volvo station wagon was parked. It was gone. Today, no one was home. I looked leftward out of the kitchen window to the other neighbors. A white house, and apparently no one lived there. Silence. Total silence. There was a blue sky and that made me feel okay for a moment as I thought about how it would be great to play some sports today. Then there was the thought that there would never be any sports again in my life.

I wonder what Mark did on these days in Fairfield. He had something always going on, at least when we were kids. I went over to the telephone in the kitchen, the white phone that always sat there silently, and picked it up. "Mark", what is going on today?" I asked.

"I am getting ready to take out my Cessna 172 on a short trip," he said.

"Can I come along?"

"What for?"

"Come on, Mark!" I whined.

"Okay," he said without adding "spook, spic, spade, gook" but that is what I heard anyway. I hung up the phone and the boredom and loneliness was gone.

Mark's house was on a cul-de-sac, but it was a brand new house and there were no other houses even built yet. I knew the way there—having been there before on one drinking occasion. The last few turns through the residential neighborhood were eerie. There were no houses whatsoever, just empty lots and real estate signs. There were no trees. Just brand new asphalt streets. First, I made a left turn on to Cedar Avenue and then another left on to Lark lane, down three empty, barren blocks to Oak Court. At the end of Oak Court, my childhood friend lived. I pulled my Dodge into his driveway next to his brown, dented, and rusted Honda Civic. He was standing at the top of the driveway and said, "Careful—don't scratch the paint," as I walked past his car.

The Cessna he was referring to had a wingspan of about four feet and was radio controlled. Painted blue, it shined as if it were perfectly new. There was a yellow strip going along the fuselage on one side and a red stripe on the other side. The bottom of the plane was painted yellow. He pointed out that when the thing is 'way up there, you really did not know if you were flying it upside down or not, and the different colors helped you know that." He said that the last time he had it out flying, he had "stupidly" painted both the top and the bottom of the plane blue. Then he got into a position where the plane was flying upside down and he "flew the thing right into the ground." "It took me over a month to put all the pieces back together and repaint it." This would be the maiden voyage of the rebuilt Cessna.

We got into his Honda and drove about two miles to a large undeveloped field—a place with no power wires—and he got the plane out and began to fuel it up. The smell of that fuel was horrible, like the worst mixture of rotten kitchen scraps plus gasoline plus God knows what. A small battery was needed to get the thing going, along with a tiny starter motor that Mark held in his hand. It came to life and was quite loud about it all. In fact, I plugged my ears, noticing that Mark had earplugs along with him. The fact that he did not bring any plugs for me was something I should have paid more attention to. At the time, I blamed myself for calling him and inviting myself at the last minute.

Mark was quite skilled at commanding the aircraft; he had it doing loops and rolls immediately upon takeoff. It was quite something to see. He made an approach to the landing strip (a dirt road) and then pulled out of the approach, heading almost vertical it seemed, and then spinning clockwise and going gracefully flat with the airplane again. There was another approach to the landing strip, and a successful landing. At that point, he just said, "I think we should leave. I have some other things to do." I questioned why we were leaving so soon, but instead took his statements at face value. We loaded the impressive plane back in the Honda and departed.

When the car started, he asked, "Have you heard that Bruce was in the hospital?"

"No," I replied.

"He says he was in a hotel in Colorado Springs and some guys just came out of the middle of nowhere and jumped him."

"No, why would they do that?"

"He says it was for no reason whatsoever."

"How bad is it?"

"He has a fractured femur and pelvis and an concussion."

I imagined that he was in pain and I was horrified. "Which hospital is he in?" Mark was pulling into his driveway.

"He is in a hospital in Colorado Springs."

"Have you visited him?"

"No."

"Why not?"

"Shut your mouth and mind your own business, n—" came out of Mark's mouth. He could not bring himself to complete the n word in full.

"Get out!" he commanded. I stood in the driveway while he took the plane out of the back of the car. He pushed the button on the remote garage opener and we walked in, closing the door behind us. The plane has its own place on the wall, a set of hooks set about two feet apart, with soft rubber covering, so the plane could be efficiently stored out of the way without taking up too much room.

"Let me show you something," he said, leading me into the main part of the house through a small door. We went back into one of the bedrooms, one that was apparently dedicated exclusively to weaponry. It has a table in the middle, with many shell casings and a reloading press. A reloading press is like a drill press, except that it is used to push the lead bullets into the brass casings so they can be reused. He pointed at the table and said, "I am glad to be able to start shooting again. That stuff was all put away for way too many years." He went over to the closet and opened the white wooden door. Behind the door was a large brown gun safe with a black combination dial. "Don't look while I unlock this thing," he said. So I turned away. When I heard the door unlock and open, I turned around to see a large collection of pistols. There must have been twenty.

"This is the one I want you to see." He grabbed one particular item, which was inside its own black plastic case. He snapped open the lid and out came a gold gun. "Remember James Bond?" he asked. "This is the same Beretta that they used for the movie.—a 9 mm model GCH1660." I dutifully said that I appreciated the pistol very much.

"It's beautiful," I said like a boy looking at a girl. It was probably the wrong choice of words. I was being too nice.

"Here, look at this," he said. He spun around with the gun in his hand and shot the thing. "BAM!" It was ear-shattering. I couldn't believe what he had done. Then I noticed that he had not taken his earplugs out from the trip. I put my hands over my ears. "God damn it, Mark," I said, "my fucking ears!" I walked over to where he had pointed the thing. He had stacked ten alternating layers of bricks and pillows in a wooden box in the corner of the room. There was more than one bullet hole in the assembly. We peeled off the first layer, which was a pillow. The second layer's bricks were all broken into pieces, mostly but were contained by the wooden enclosure, which had separate compartments for each layer. It worked. He pointed out that there was simply no chance of a ricochet with the design he had used. Apparently, he was still reading *Soldier of Fortune* magazine, where he got all these ideas.

I continued to admire the layered backstop, and wondered if that kind of thing is what everyone in the town of Fairfield is doing—shooting guns indoors on a Saturday. I scanned my eyes across the white painted walls and there were no decorations or wall hangings of any kind. There were heavy drapes across the window, so I pushed them aside. The window had been papered over from the inside. "I don't like curious people," said Mark.

"I guess not," I said, feeling bewildered. I had come over here to shake off some boredom and wondered what kind of alternative this was. I walked out of the bedroom into the hallway and turned left, down the hallway into the kitchen. Mark was behind me and said, "I'd offer you a beer but I know you don't drink anymore . . . how about a glass of water."

I said, "No thanks." My ears were still ringing.

He sat down in a chair at the kitchen table, a little plastic table he probably got at a thrift store. "Do you know that guy that Bruce works for, Don Getz?" he asked.

"Yeah, I met him the day I arrived at this place," I said ironically. (We were not at the lab, we were at home.)

"Do you know that he cancelled the program my friend Mike West was working on and sent everyone off packing?" I said nothing. It seemed to me that this kind of thing happened all the time.

"So?" I said.

"He had everyone go off looking for new jobs, while there was still a million dollars in the budget. Then, after they had all been reassigned, he spent the money on hiring five new post-doctoral researchers working on the C-Charm project."

"C-Charm?" I asked.

"That's the one where they are measuring charm quarks at the Brookhaven laboratory. Those positions require travel back east to the experimental facility for a year at a time, so each post-doc costs a quarter million dollars a year." He said Mike West had spent several years of his own time bringing that funding in through an Air Force contract and was told to stop spending the money immediately. "Now, Mike is in Earth Science

Division doing seismology work. That is about as far away as you can get from where he got his doctorate, in laser physics."

I really did not understand. I figured that managers are paid to make these important decisions and that is what they do and it is not up to us to second guess what they are doing. It was probably for the best. I know that I had my differences with Don during the job interview, but that didn't mean that I should automatically believe he is some sort of crook.

"Stan, you must understand, there are some people who simply have no conscience."

"By that you mean Don Getz."

Mark affirmed, "Don is a person who has never known any guilt, remorse, or regret for anything that he has ever done. He is completely without moral scruples. The only thing that matters to him is what is best for him and his cronies."

While Mark said the word "Cronies," it occurred to me that *Bruce* was Getz's crony. Furthermore, the director of the laboratory appointed Getz to his position. The director was a highly respected scientist who had given public talks to the Commonwealth Club, to National Public Radio, who had testified in front of Congress. The Fairfield lab was considered one of America's "Crown Jewels," according to the secretary of energy, who was a presidential appointee. I was not mistaken. This was truly a great institution, according to the press releases.

Yet my friend sat in that chair in front of me and said that years of Mike West's work could disappear at the whim of a crooked manager.

"Is that right?" I said incredulously.

"It sure is," said Mark.

"What does Mike West say now?" I asked.

"Mike is just hanging on until something better comes along."

"I had better go now," I said, starting to feel bad. My ears were still ringing like hell. I crossed the kitchen and headed towards the front door. On my left in the hallway was a small drawing framed. It was a picture of a cat with a bull's eye drawn around it.

Fairfield Career

Starting at Fairfield Lab, I was assigned to Nancy Jordan's group building laser diagnostic equipment for hydronuclear experiments. This is basically a test or experiment to build a nuclear bomb that only goes bang instead of boom. Then you see what would've happened if it'd really gone boom by doing some computer simulations.

Nancy was a chemistry PhD from University of Arizona. She had spent most of her career in the debris transport field, making debris maps from models of large events such as the Chernobyl accident, and a similar British nuclear accident from the 1950s called Winscale. Although she was not blonde, I still noticed her good looks and, with disappointment, that she was married.

Mark's group had dissolved and there were several PhDs looking for jobs. At the same time, P division, and the advanced warhead group in particular, were growing in size and in budget. They had apparently landed a new project that was reported in the Fairfield newspapers as the W911. It was also known by its nickname, ATUM, the "Advanced Tunneling Underground Munition."

To me, it simply seemed beyond belief that a new warhead would be designed so many years after the Cold War had ended, but lo and behold, there was a new nuke. The local townsfolk were chattering and the Oakland TV station carried a story on it. It seemed that there was a postulated threat of biological weapons facilities buried deeply underground in "rogue" nations. This ATUM could penetrate the earth nearby and actually tunnel its way down 100, even 200 or 300 feet underground before detonation. The blast yield on this item was tiny by nuclear standards, only 10 tons, with a correspondingly low release of radiation. The claim was that the yield was so small and the penetration so deep it would not release any radiation from the underground cavity where it detonated. It would be a nuke we could actually use if we had to.

I read the article over again and again and was bewildered and amazed by this thing. It seemed to me to be too good to be true. Why, we could knock out everything they had stored underground in North Korea without any fallout on South Korea or Japan. It also worried me somewhat in ways that I could not really put into words at that time. It was probably the idea of actually using nuclear weapons.

I got Mark on the phone—he was part of the underground experimental testing for this postulated device. Bruce was section head for the theoretical analysis of the ATUM penetrator calculations and was heading up an elite team from Princeton and Caltech doing full three-dimensional physics estimates using a large supercomputer. The computer manufacturer, Intel, was providing a large cluster of 7000 nodes running at 3 GHz for almost $200 million. Bruce's wife was on the Fairfield team overseeing the contract. Mark had a girlfriend again and she was working for Intel.

"We should get together sometime for lunch," I said, but it felt pretty lame as soon as I said it. I thought to

myself "*Look* at us doing lunch now—what a bore."
Then I said, "How about lunch and a refrigerator?"
There were some muted chuckles. "How's the Beretta?"
No answer.

There was also to be no lunch, but we kept talking
on the phone. I had an idea for something much more
interesting, or so I thought and brought it up. Mark was
in Steve Hurlbut's group and would talk to Steve about
allowing me to give a guest lecture on the pluses and
minuses of the ATUM.

"What was that?" Mark said.

"I said plusses and minuses."

"What the hell is this minus shit?" he shot back.

I explained to him how the physicist Albert
Wohlstetter had explained to President Kennedy years
ago the problem with low radiation warheads. The
problem is that they may actually be used. What had
been keeping humanity alive is that these nuclear
weapons are too horrible to be used. But the calculus
can change somewhat when a warhead becomes
available that has no radioactive fallout. There's a
lower threshold for their use.

There was a click on the telephone. He had hung up.
Certainly, this was some sort of mistake. He didn't
understand the concept of balance and dialogue. Getting
to common ground, etc. I called him right back. Out
came the F word, the N word, the S word, and then he
hung up again.

A call to his supervisor would do the trick, I
thought. Steve knew me somewhat from Berkeley and
we had always gotten along. I explained to him the

need to have an open dialogue about the pluses and minuses of this new warhead. I can facilitate the discussion by giving this talk. It would be all very interesting to everyone involved. "The regular engineering test design group meeting is scheduled every Tuesday at 7:30 AM and there is no speaker scheduled for next Tuesday."

Deterrence Theory

We had a plan. I had a plan. I would go back to Berkeley and check out some old books that I had read before. I could start with the Hiroshima story, moved to the SALT-I and SALT-II treaties, the theory of deterrence by Tom Shelling, and transition through the proceeds of the Reykjavik meeting. It was going to take all weekend. More than that, I was taking Friday off to go to the Bancroft Library and gather my wits.

The weekend went as planned, and when I awoke on Tuesday morning, there was an incredible rush of adrenaline. My heart was pounding—I was so looking forward to this talk. I woke up at 5:00 AM and began reading over my material. I must've had twenty cups of coffee. When I arrived at the room where the meeting was to be held, only Steve and Mark were there.

"What happened?" I asked.

There was a slight delay, as if they did not understand the question.

"We are here—isn't this what you want?"

"Where is everyone else?"

"There is no one else except Ronnie Boyd and he is on travel out of state. So let's hear your talk."

I gave the talk as if nothing was the matter, launching into the entire thing as planned without holding back on hand gestures, oratory, or hyperbole. Upon showing the final view graph, I paused dramatically before speaking any further.

"As far as we know, the ATUM is an advanced warhead that adds little or no security but increases our dangers tremendously, especially if the technology were to get into the wrong hands."

Steve and Mark were silent. In fact, you could say the silence was deafening. Neither one of them asked questions, but Steve thanked me for the fine talk. So I walked into the hallway down the corridor and out into the parking lot.

It was still before 9:00 AM. Some people were still showing up for work. I thought, *How nice it is, they drive such nice cars like BMW, Lexus, and even Mercedes-Benz . . . all on public dollars.*

Bruce's office was diagonal across the parking lot, in the second floor of the new building named the ATUM Center. Oh, boy, what am I getting into? Walking upstairs, it was heartening to see that name plaque Kozumplick. His secretary blocked my path to his door.

"I'm here to see Bruce," I said, trying to sound like I belonged there.

I had worn my suit and tie that day to work because of the talk.

"Who should I say is here to see him?"

I could not crack a smile or even tell a silly joke. "Stan Hall."

She said, "Bruce's in a meeting now. You can see them at 2:00 PM today. I'll put a meeting on his calendar; how long do you think you'll need?"

"Oh, about an hour," I said.

I came back at 2:00 PM and Bruce opened his door to me. I had not really seen him face-to-face in at least a full year.

"Well, little Stanly, what can I do for you?" he said with a smile.

"Bruce, I just wanted to talk to you about the ATUM program. It seems like it can do more harm than good."

Abruptly, he said, "I've heard all that before. The program is one of our country's greatest needs, and it is growing."

"But what if someone else like India gets this technology? They might actually use it," I said.

"You don't know what you're talking about—we have 400 of the best theoretical physicists in the world and a half dozen computers carrying the capability of 1000 Cray-2 machines each, performing all these calculations. There is no country on earth that can ever do that."

He took a breath of air, apparently so he could continue speaking as quickly and loudly as possible.

"Our next set of Intel chips will arrive soon with quad cores and 3 GHz speed and over 70,000 nodes with 10 processors per node."

He showed a chart of how a variety of supercomputers rank against each other in terms of computing capacity, location, user, owner, and the year first used. Fairfield was no longer in the number one position, but it would be after the new Intel cluster arrived. The chart said there would be 7000 nodes with 10 processors per node. I did not call this speaking error to his attention.

His smile returned, but his face was still tight. "So go rethink this crap going through your brain." He had delivered his message, a message that had some inconsistencies, but I knew it would be futile to discuss those shortcomings with him. He had a certain posture and pace to his speech that indicated he had been through this all many times before, and that I certainly had nothing new to say that he had not heard before.

I tried a change of subject. "Bruce, I want to say thank you for driving me and Rudy all the way home from the baylands that night. You saved us a really long walk." It was not immediately clear to him whether I was serious or pulling his leg. I was hoping for a chuckle.

Instead, he got very angry and moved a foot closer to me. "You and him are the two stupidest kids I have ever met. I should never have had anything to do with either of you."

Just then, his secretary walked in. "Your reservations for Japan are in. Here are the tickets—I got you window seats going in both ways."

Taken aback, I said in a polite tone, "What is going on in Japan that you need to be there for?"

He said there was a nuclear physics meeting, only being polite and answering my question because his secretary was still there.

"Okay, Bruce, sorry to bother you," I said, easing my way out the door.

Looking up at the wall clock, it read 2:20—only twenty minutes had gone by. As I walked down the stairway to the first floor, the thought occurred to me that nuclear physics was no part of the atom project. Besides, you could not hold a nuclear weapons conference in Japan under any circumstances because classified conversations are not allowed to take place in Japan (and the Japanese would not allow it anyway).

Back in my office, I looked in my most recent copy of the technical magazine *Physics Today* for the conference and found it scheduled for Kyoto, lasting two weeks, starting about a month from now. I brought up the program listing. Bruce's name appeared on a talk entitled, "The quark spin state, yesterday, today, and tomorrow." I knew that a quark conference was the last thing the ATUM would ever need.

After a few months, our hydronuclear program faded in importance and I was in somewhat of a bind for work. The ATUM program was hiring people to build computer code for the new Intel cluster; the one that the newspapers claimed was now the top supercomputer in the world. I had done some theory as a student, and the idea of going back to theoretical and computational work was okay to me. The division where Bruce and Mark both worked had developed a reputation as a truly good employer, with many performance awards posted in very public places throughout their buildings. Don Getz was still the division leader and Bruce the deputy. Mark had been promoted to team leader, after Steve had left the lab and moved over to Silicon Valley.

ATUM Instability

I made the move to the ATUM program. The task was to develop, or help develop fluid models showing the trajectory of the warhead within the soil. An issue it opened was whether it would go in a straight line as intended when underground, or whether it would veer to the side. It was said that the tendency to move sideways was increased when the angle of the warhead entering the ground was not perfectly straight up-and-down. The team had already been working on this problem for a full year and we were converting its coding system for the new Intel cluster.

At the first meeting I attended, about a dozen engineers showed up and introduced themselves. Team components and functions were explained. Tom Christiansen was the one who did the most explaining—he had been with the team the longest. The focus of his work was the so-called Rayleigh-Taylor instability at the surface. This instability, or anomaly, would cause angular departures to be intensified. In other words, if the nuke was not going straight down at the start, any departure would be amplified, making the nuke go sideways and then possibly turn around and burst through the surface again.

I was to work with that team for a full-year before discovering that the Rayleigh-Taylor instability, when properly understood, would also guarantee the warhead from traveling any more than about 20 feet downwards into the ground. After that, the warhead would always go sideways.

This could be explained by thinking of a pool of stagnant water, where a round aluminum plate is carefully made to rest flat on the surface. It is thin, perfectly flat, and denser than water, so the plate does sink into the water. A question one may ask is whether the plate can go straight down to the bottom of the pool. The correct answer is that it could only go straight down to the bottom of the pool if one were to neglect all instabilities. Otherwise, the plate will tip to one side and shoot down sideways at an angle to the bottom of the pool. In fact, this always happens. The plate cannot go straight down.

The problem is far more complicated for a warhead penetrating the soil. A complete explanation would take hours. The idea of giving a talk on the subject came into my head on a Friday night. A forum on Monday at 7:30 AM would be the right place to discuss this topic. "RT instability applied to the ATUM depth problem" was the title and I was on fire once again. By 9:00 PM, there was an outline, by 11:00 PM a few diagrams, and a few hours later, text and labels for the graphs.

At about 3:00 AM, I remembered how much fun it had been to blow up the refrigerator at Ford Aerospace and could not help but laugh hysterically. I looked into the mirror while laughing and my entire face was red, but the huge scar on my left cheek was light pink. The whites of my eyes were actually deep red. My eyes wandered around looking at all the defects. I said into the mirror "Stan, you are still ugly." There were wrinkles, blemishes, and now a receding hairline addition to the scar.

"At least I don't drink anymore," I said to myself.

Standing back away from the mirror, I saw that I had regained all of the weight I had lost "over Christina" and had gained another 30 pounds beyond that. The scar, which had been with me for years, seemed like a confirmation of something. I was an ugly man, but at least I could give this talk, and it seemed so important.

The piles of paper all over my house in Fairfield, which I had taken with me from Berkeley, told the story of many projects that had been started, then thought of as the most important development of my life, and then had been canceled. My eyes moved from one pile to the next, memories flashed for each piece of work. None of them had ever been real. It had all been make-believe.

At 4:00 AM, there was nothing else to do but fall asleep. Falling asleep was the only real challenge of the night as I tossed the covers off and wrestled with the pillows for an hour at least.

At the 9:00 morning forum, there was more than the usual number of folks in attendance. To my great surprise, Bruce was there, too. Mark came in wearing a set of leather motorcycle riding gear.

The talk went well. Tom and the other engineers seem to be feeding me the right questions. The only difficulty came towards the end, when the connection between theory and experiment had to be explained. There were not enough experiments to show exactly how the departures along the predicted theoretical line would take place. I proposed doing these experiments as the final part of my talk. Tom and the rest of the team were enthusiastic. Bruce was silent and looked at the floor. Mark held his helmet in his hand for the entire talk until that point. Then he turned towards the door and put it back on his head, and the door closed behind him. I felt a tinge of fear at his departure for some reason.

The rest of the audience was chattering with each other. Bruce spoke to the person next to him for a moment, excused himself, and then exited the building. When Bruce left, I no longer cared what anyone else was saying.

Tom followed me to my office and we went over the ideas in the talk again, forming a plan to propose a set of experiments. There were several people who had to be involved. A budget had to be developed. Some people from other divisions had to be involved. Tom and I surely would have a proposal completed within a few weeks.

I made a phone call to Mark from my office to ask him why he had left early.

"Stan, you don't really know what you're doing."

Certainly, he had misunderstood me, I said.

"We can't talk about this on the phone," he said.

Of course, he was right. Some of our conversation would contain secret information in subject areas outlined in our division guidance as never to be spoken about over the telephone lines. Classified conversations are to be carried out only behind closed doors.

He would meet me face to face later in the day, so I tried to arrange a meeting.

"I am 'way too busy today—my whole day is booked."

"But I want you to be on this project," I said.

"There is no project," he said.

"Okay, I mean proposed project."

"The project will never see the light of day."

After I hung up the phone, my office was a haunting place. Darkness and chills went up my spine. I closed the door and my stomach started fluttering again, as it had done years before in the apartment in Berkeley. I looked at the clock and it was 11:30. Then I looked down at the floor. Then I looked back at the clock and it was twelve noon. I had a feeling of utter aloneness again, as if I was the sole survivor of the human race after all life on the planet had been destroyed.

I thought of Christina and how she had once been the only thing in my life that meant anything. It seemed a little humorous now, and that thought gave me a little chuckle. The entire episode with Christina had all been for nothing. Then I thought about how I had worked so hard for those years in Berkeley, and that I had ended up in a place where new ideas are actively hunted down and destroyed.

Quarks

Bruce, I thought, hated me, and that is why he had walked out. Then there was his boss, Don Getz, who hated me even more. I have to speak with at least one of them and try to make peace. Bruce or boss? Bruce it was. Again, I walked across the parking lot, with my solar plexus muscles fluttering. "Stop it," I commanded the muscles. Up the stairs to the second floor, his secretary was out for lunch and the door to his office was closed. A sentence fragment came through the door.

"No more budget . . ." the voice sounded like Getz's voice.

"We'll get through the next conference without falling short," I heard Bruce say.

He must've been referring to the upcoming nuclear weapons conference in Los Alamos, the one that both Tom and I were attending.

"Speakers include both Woljeck and Gell Mann."

He had dropped the names of two Nobel Prize winners. It was a quark conference.

"We have sixty-five people going to Switzerland next month and the director wants an explanation in writing," said Don.

"We were light at the last conference, we were also light on the Higgs conference two years ago."

"Different budgets," says boss.

"We can use carryover on NW 227."

"Brilliant!" laughs boss.

My ears cannot believe what they are hearing. There are sixty-five people attending a quark conference in Switzerland while charging their time to the NW 227 budget. That was the budget code for the ATUM penetrator calculations.

The next topic comes up.

"What about your fucking buddy?" says Getz.

"We cannot fire someone for having their own scientific opinion," Bruce said in a firm voice.

"You hired that bastard," says Getz and there is a tense chuckle.

"We'll see what we can do," says Bruce.

It sounded like the conversation was ending and I had a strong impulse to run away quickly. My fat body bounced down the stairs to the first floor and to the nearest exit. As I trudged across the parking lot, the old irony of the ATUM building changed to a new irony. They knew that the warhead would not even function as advertised, and they did not care in the slightest.

When I had first arrived, I was terrified that the ATUM was an unnecessary tool—potentially dangerous in the wrong hands. It needed to be thoroughly evaluated for its utility. My entire thought system had been based on the assumption that it actually worked. Now, I realized that nothing at all was ever to come out of this program, except a nice building and an Intel cluster. It was all a way of providing high salaries for people like Bruce to play at science and travel to conferences. It provided jobs for people like Tom, Steve, Mark, and me. It provided us with something to do with our time writing computer codes for the large supercomputers. The only thing that really counted was that our adversaries, those rogue nations building underground facilities, believed that it would work and could be used. The fact that it could not work was hidden by secrecy rules, and the fact that we would never really test the system.

Before even sitting down in my office chair, I felt like crying. But I could not cry. This was a nuclear weapons lab, the door was open, and someone would hear me. I looked at the phone directory, available on the Fairfield web page. One time, when I was first hired into P-division, Don Getz had told me that all matters of security and workforce abuse were to be handled in the operational security (or OS) division. Looking up "OS," it appeared they had recently changed their name to "AA" which stands for Auditing and Assessment. I could not believe I was going to turn in my friend, or former friend, Bruce, but I simply had to.

An appointment was made for the following Monday with an investigator, Carmen Roybal, who would handle my case. She asked me to fill out a one-page form about the nature of the problem, in advance, so I typed in the details the best I could without revealing classified information in an unclassified form. Carmen did have a clearance and we met behind closed doors in an office near the laboratory director's office. Certainly, this was the right thing to do. She explained to me that AA division would be tracking this issue closely and as quickly as they possibly could.

"In the meantime, we recommend that you continue working on your job, performing your duties, and not causing any further disruptions."

I smiled and thanked her. After I departed her office, it occurred to me that she might have been using the word "disruption" to refer to the Rayleigh-Taylor proposal that I was writing. Is that what she meant? What a strange choice of words, I thought.

Tom and I continued to work on the proposal. I tried to have a meeting with Getz more than once about the proposal, but could not get on his calendar. Bruce seemed to be on travel roughly 99 percent of the time, sometimes to Switzerland, or Russia, or Japan or sometimes just to Washington DC.

Seeing as not a word had come out of the AA office, I scheduled a meeting with the head of AA Division, John Mayer, who was in the office immediately next to the lab director. The meeting took place on a Monday morning at 8:00 AM, and he seemed to be a little apprehensive when I arrived. I explained to him that I had seen a case of sixty-five staff members traveling for two weeks to attend a quark conference in Switzerland all on the ATUM budget. I also suspected that the ATUM would not even work as advertised. In the meantime, I had also found there was no funding to complete some crucial experimental work. How could this be?

"We asked you to continue working at your position and perform your duties," he said.

There was silence from me—I thought we were going to discuss the corruption within the ATUM program.

Then he said, "Did we or did we not tell you to continue to perform your duties?"

"Yes, but it is in my job description to write proposals. That is what I have been doing."

He indicated that my job description included designing computer experiments that showed "a linear trajectory for the warhead while under the soil." He said that I had been computing nonlinear trajectories, instead of linear ones as required by my job description. Further, the proposal I was working on was for nonlinear trajectories only, which is also a violation of my job description. His face was twisted with anger and outrage while he was telling me this.

I was speechless. I was dumbfounded first of all, that he had actually read my job description. Secondly, that he had wildly misinterpreted my job description in a way that made my actions on the job a violation of that job description. What he was telling me was that I was only to bring forward studies that confirmed the ATUM would work only as advertised.

"I am a scientist," I said.

"No you are not," was his reply. "You are here to do what we tell you to do."

"But what about all the staffers on travel for Swiss conference who charge their time to ATUM?"

"That is none of your business."

Silence.

"You call yourself a scientist but you're really just a worm," he said. "That place you came from, Berkeley, we built that. It all came from the nuclear weapons budget." My breathing became elevated. "And your advisor, Prauznitz. He got his start all because of the budget we provided him as a post-doc. All that chemical engineering research is just a bunch of bullshit anyway. You would be nothing—absolutely nothing—without us. Now, here you are complaining, biting the hand that feeds you, like a sick animal. Why the hell don't you grow up?"

I looked at him in the face, clenched my fists together, and said "You Goddam bastard!" I looked up at his wall while I was leaving and he had a Cornell University physics PhD diploma prominently displayed. Yes, it was ironic. He was someone who had spent a great deal of effort to obtain an advanced degree from a prestigious school, ending up in a position of concealing waste, fraud and abuse in a thoroughly corrupt organization.

The Money Bomb

Instead of going back to my office I went to Bruce's instead. His secretary advised me he was on travel to Moscow.

"Does he have a phone with him?"

She gave me an international cell phone number, and so I looked at the wall clock, counted eleven time zones to where he was. It was 10:00 PM there. I went to my office and called him.

"Bruce, remember the great times we had at the Stanford pool?"

Silence.

I continued, "It sure is interesting how well the ATUM program is going."

Silence. We both knew that the conversation was being listened-to by the Russians. Besides, what could he say?

"Bruce, you know what I know about the ATUM, don't you?" I thought he would hang up, but he didn't.

He finally answered, in a firm, forceful, yet calm voice, "Stan, it's all a game. Just play the game."

I had certainly not anticipated that response. Without directly saying it, he had confessed. None of this ATUM project was real; it was all simply about money. This confession left me speechless; I just hung up. His advice and his tone of voice "Just play the game," sounded eerily like the encouraging words of an older sibling.

He used the word "game"—like the prisoner's dilemma. If we all stayed silent, the paychecks would keep coming. At Fairfield, there are more than just two prisoners in the cell. It is not quite as simple as the prisoner's dilemma. There are many, many players in this game. Hundreds of people have jobs playing with supercomputers. Even Intel Corporation is involved, because this is a way of the US government subsidizing the US microchip industry. The entire town of Fairfield and all its occupants benefit tremendously from this game. All benefit and all stay silent.

From Bruce's perspective, the game was "I keep the paychecks coming and you let me do what I want to do."

Someone else at Fairfield had once said that the place was just a sandbox for physicists to play in. But I thought that the play was serious play. After all, it seemed only a few were "playing." Many of us were really working. The funding comes through a set of programs and our program manager at least seems to care what happens. For this game to exist, the program managers must not know about it. Or do they? I wondered.

The game I knew how to play was a combination of mathematical analysis and experimental study. That is the very elementary and primitive game called science, the one I learned in school. It is such a simple game to understand compared to the one I was in. But how do I play the Fairfield game?

From what I had seen and heard in John Mayer's office, it was clear that there would be no turning back. The study proposal would go forward as planned. Tom headed the experimental side and I continued with the theory. Computational resources were found, and would be made available if we had the money to pay for them. The experimental apparatus and manpower were available, also if we had the funding.

A visit to the program manager seemed like necessary step in finding out about this game. Not the game as Tom and I played it, which dealt with math and science plus responsibility. Robbery was the game that was played by Bruce and Don behind closed doors. I wanted to know more.

Doug Stewart had been at Fairfield, it seemed, since it was nothing but a cow pasture. He had a remarkably large office with the conference room essentially built into it. I asked about his background and he affirmed that he was one of the few who actually had been in attendance for an aboveground nuclear test. He had then stayed on in J division (nuclear testing) for two decades, supervising a series of underground test shots. He later was the head of computer simulations for advanced stockpile weapons. He was happy to tell me about the two years he spent as the Fairfield adviser to the Air Force on nuclear weapons use doctrine.

"You mean they actually have plans to use these things?" I gave a deadpan face. I really wanted to know what he had to say.

"Hell, yes," he said, smiling.

"Against who?"

"Well, in the days of the Cold War, the use was against a set of targets in the Soviet Union."

"Certainly you know," I said, "that the only purpose of these things is only to deter the use by the other side." I felt an ironic smile come to my lips. I could feel a little twist in my jaw. "What would be the purpose of actually using these things?"

He looked at me with a smile and said, "You and I may see things this way, but they do not. To them, the purpose of war is to win the war. They have their orders and they will carry them out. There is to be no hesitation."

I thought for a second. "What about nowadays?"

"Nothing has changed, not really," he replied.

He knew about the problem I was working on and said that people at Fairfield had brought it up a long time ago and had dismissed it. "Back then, we did not have the computational resources to do the full simulations, so we simply dropped the subject.

"We now have the computational capability to actually get some real answers," I said.

"You know that you will have to get those machines away from the people who think they are entitled to them—that is not going to happen."

I became a little angry with my new friend.

"But…," I started.

He interrupted. "Every young red-hot Berkeley graduate who comes into this place wants to do the same thing—change Fairfield." He then continued. "But they won't let it happen."

"Who is 'they'?" I asked. "After all, you are the program manager."

"I just help keep the paychecks coming. The computational facilities were bought and paid for by a deal worked out by Ray Yosky with the Air Force and the DOE back ten years ago. They will not let you use them to disprove the validity of their pet project, their cash cow."

The little meeting in his office was over because I was running out of things to say. Nothing seemed to startle him—he had heard it all before. During my walk across the lawn in front of his building, I broke into laughter. "Cash cow," as in something you milk for paychecks? So you can continue to play in your "sandbox?" The laughter came out loudly in a few moments. Followed by wincing. For some reason, I found myself angry beyond circumstance. To this day, I do not know why I did not just shrug it all off.

It seemed peaceful at Fairfield, as if I were the only angry person there. The day ended early again; I went home and watched TV. The good guys in the *Burn Notice* rerun, Michael and Sam, put some bad drug dealers away, only to have them let go because they were valuable to the CIA. But that was just television. The TV had a commercial break and an announcement that at 10 PM the evening news would cover a protest being held outside of Fairfield Lab. An old hippie woman held up a sign saying "save our children." She probably meant something about children being killed by bombs. But what is really going to kill our children is debt. "A Money bomb," I thought to myself, "that's what Fairfield is." Another chuckle, muted.

The only hand to play in the game was to try to get computational access on the cluster for my calculations. The final proposal was put together, starting on a Friday afternoon, requiring all weekend to write and rewrite. I called Tom at home on Sunday afternoon and he agreed to meet first thing Monday to clean it up and send it in. This is despite all the warnings I had been given. This was a proposal to try to disprove the central hypothesis of a major weapon system using the resources of the system itself.

Monday afternoon, before the proposal was submitted, I went to see Mark to find whether he wanted to be part of the project.

"Why would I want to commit career suicide? I've worked hard all those years getting through my master's degree. It was a long path getting here." Then he got angry. "What the hell would this accomplish?"

Out of my mouth came, "This could be the start of a new program for a non-earth penetrating weapon," I said. But to myself, I thought, *I do not really know*.

"The entire project is a fraud," I said.

"Congress voted for it," he said.

"Mark, you don't understand. Those thousands and thousands of computer processors are actually churning out junk numbers. Only about half of the theoretical physicists on the program or about 100 people are actually doing the work they are supposed to be doing. The other half is completely unaccountable. This is on a project that costs $400,000 per employee per year. One hundred staff members mischarging their time for a year is $40 million dollars a year."

"What am I supposed to do about it?" he said.

"Mark, someone has to do something."

He said, "I'm too busy to talk about this anymore."

He stood up, still wearing his leather jacket from his morning ride, and walked me to the door.

A cascade of bad feelings came over me. He had been my friend. "Once you get out of graduate school, it's just dog-eat-dog?" I asked myself. I walked back to my office and put the proposal into the computerized system, after getting approval from my line manager. In a state of denial, I called Mark on the phone.

"Mark, I have a flyer here from the inspector general of the Department of Energy with an 800 number to call."

Silence.

"I will report only that there is a set of fraudulent charges in that sixty-five staff members are traveling to a quark conference while charging time to their weapons project."

He replied, "I can see why you're mad." Then he said, "People like us have to work our butts off twice as hard because half the people are doing completely irrelevant things."

He was right, although I hadn't thought of it that way until then. Maybe I had gotten through to him.

"Mark, do you know who is the head of this whole scheme?" I asked.

Silence.

"It's Bruce."

"What do you mean?" he asked.

I told him about what I had heard through the closed door. All of Bruce's time charging fraud was being conducted not only with his knowledge but he considered himself the chief in charge of planning for all those false time charges. Then I said, "Mark, he told me it's all a game."

Mark had a few words of his own to add at that point.

"Bruce and I used to be friends long ago. He used to let me ride in the first car he ever owned."

Stole, I thought to myself.

"Something happened in the last year of high school when he got back from Montana. He seemed like a different person—he was using hard drugs, he beat Mom up—that's why he was thrown out of the house. Now, I see him here and he just looks right past me like I am not even a person, let alone his brother."

I was thinking that's basically the same Bruce I'd known. He would completely ignore me unless I got in his face. It is as if we had never known each other before.

Mark continued, "He works with that guy, Don Getz. Getz is a person who has no conscience whatsoever. The man just spits venom at anyone at all who is either not someone he wants money from or who is not his puppy dog. I have tried to speak to him more than once and all he could do was condescend to me. Then he started yelling. He is a pure asshole." Mark went silent for a few moments. Then he continued, "The few times I've ever spoken with Bruce, he treats me like . . . he treats me the same way. I don't know if he has gotten lessons from his master, or what."

The IG

The phone conversation ended after I had gotten Mark's agreement that he would testify to the inspector general. I then took my leaflet and walked over to the program manager's office.

"Doug, do you realize that the ATUM team is just milking your program for whatever funding they want and manipulating the program as it suits their desires, and just laughing at this thing behind your back?"

"Yes, I know, we already talked about that, but what can I do about it?"

Then I pulled out the leaflet. "Doug, I'm going to be calling this 800-number. I have personally witnessed sixty-five staff members charge their time and travel to a quark conference in Switzerland and I will testify to that. I will give them your name and phone number and you can speak to them, too."

"Stan, you are just a young guy here. They will do whatever they can to ruin you."

I looked at him in the eyes and said, "Okay."

He said, "You realize they will make sure you are never allowed to do anything here again?"

I was silent. I must have been giving my silent consent.

He said, "I'm an old man, and I'm retiring in two months. You came at just the right time," he pointed to a box of books in the corner of his office that I had not noticed before. "I've just started packing. When I'm gone, there is nothing they can do to me."

"What are you going to do next?" I asked naïvely as if everyone went on to the next position somewhere else.

"I'm not going anywhere except Portland Oregon, to live near my two granddaughters."

The inspector general is an investigative arm of the Department of Energy, which in turn supplies Fairfield Laboratory with almost all of its funding. I had two other people testifying with me, Doug and Mark, or so I thought. The agent assigned to my case, George Mason, took my tape-recorded testimony under penalty of perjury. The scale of the suspected corruption made his eyes double in size. He was astounded at what I was telling him. I clearly stated the only piece of solid testimony I had was about the conference, which seemed like exactly what the inspector general was looking for.

He said, "You may be the eyewitness to a tiny piece of fraud, but this could uncover something much bigger."

I gave him the name of Doug and Mark and both of their phone numbers. I assured him this was indeed a much bigger problem than simply what I had seen.

Since I had waited a whole month for this agent to arrive from Washington, DC, I suspected it would take him a year to complete his investigation. But that was not to be the case. He was done in one week. I got a call from him and we met in the parking lot.

"The conference was allowed under the existing contract between the DOE and Fairfield," he said.

That had to be the last thing I expected to hear. "What about the time charges for the hours spent for sixty-five people traveling to Switzerland for a week while charging their time to a weapons project?" We were both aware that the lab charges a rate of $200 per hour per person.

"Stan, I know this sounds incredible, but the existing contract says it is up to management to verify that time charges are warranted and accurate. We can go after each and every line manager in the ATUM program, but the worst charge we can ever bring is incompetence. They can always claim they were simply approving time charges because they did not know what they were doing. Under the existing contract, no one can be fired for incompetence."

Then I asked, "But what about the division office?" thinking about Bruce and Don.

Mr. Mason then said, "It's only their responsibility to select line managers. Again, they can always claim incompetence. I'm sorry, but there is nothing we can do."

What about what Mark had to say?"

George told me he could not get a hold of Mark Kozumplick, who had gone on vacation, and was not returning his calls. George told me there was nothing he could do to force him to testify.

"What about what Doug had to say?"

George told me that although Doug had a lot to say, I was not to be granted access to that information, because of the sensitivity of the information. "In other words, you'll have to read about it in the newspapers like everyone else."

Months went by. The ATUM program continued to set records in computational throughput. The proposal I wrote with Tom Christiansen, of course, was rejected.

A position opened up on another project outside of ATUM and I put in my application. They were in need of a chemical engineer to analyze groundwater migration at the old Los Alamos site that had been active during the Manhattan Project. Groundwater plumes and smoke plumes are mathematically related, I thought. I walked across the lab thinking that this would simply be the closing of one chapter of my career and the opening of the new one. Arriving in the Earth Science division headquarters, I ran into someone I did not expect to see. That was Bruce.

"Hi Bruce," was the best I could manage.

"Hello" was his reply, devoid of other acknow-ledgement.

Entering the building, I found that the position was no longer available. Doug was right; they would not let me do anything again.

The next day, an FBI agent contacted me in my office. They escorted me to the front gate of the Fairfield complex and took my badge. I was allowed to keep my briefcase, but they searched it for some reason. A hearing for my clearance reinstatement was set for the following month.

Endgame

I appeared in Oakland in a small conference room with an agent, Judith Horning. She informed me that serious questions about my qualifications to hold a US secret clearance had been raised. I wondered what they had come up with. Maybe they knew something I didn't know.

"Stan Hall, we have reason to believe that you have falsified time charges while working on the ATUM project."

I did not know what to say, seeing that I had worked many extra hours nights and weekends, while at least 100 other ATUM team members were given permission by management to falsify their time charges.

"Have you ever charged time for one project while actually working on another?"

"No, not really," I said.

"What do you mean? Is it yes or is it no?"

"No."

"What about February 27 of this year. Our records show you charged your time to one project while writing a proposal for something called deflection simulations?"

"Oh, yes," I said, "that was the day I went to Mark to . . ." my voice trailed off. Had Mark had testified against me?

Judith said, "We have reason to believe that you've been drunk on the job. Have you ever been drunk on the job? Remember, you are under oath."

I said, "No, I quit drinking before I came to work here." It was easy to remember that day: It was a few days after passing my oral exam at Berkeley.

"When was the last time you drank?" she asked.

"December 27, 2003," I answered.

"Have you driven drunk?"

"No."

"We have a report you drove drunk all away from Fairfield to Palo Alto."

I went into a dream state. Only Mark knew about that. I changed my answer. "Yes, I did drive drunk." I had been warned by Professor Prauznitz years before, that when you speak to DOE security, to never change an answer you had given, no matter what. I realized that I had just broken that rule.

"Is there a time that you have ever failed to make child support payments."

"What?" I asked.

"Have you ever failed to support a child that you are the father of?"

"I have no children."

"We have testimony from one Karen South that you have fathered a child and have refused child-support payments despite her contacting you repeatedly for support?"

My God, they have spoken to crazy Karen. Blonde Karen. Psycho Karen. I thought how she had pointed that gun me. I thought about the screaming and the sheer terror of being near her. She got pregnant? I have a child??

After a moment of silence, I opened my mouth. "I am willing to talk to her. I am sure this is some kind of mistake."

The agent said she could not provide any contact information for Karen, whom she called a "witness." Then she asked, "Why have you refused her attempts to contact you?" I said Karen has never attempted to contact me. Judith asked once again, "Is that your final statement?" I said yes.

That ended the interview. Leaving the federal building in Oakland, I knew that I would never get my security clearance back. The game was over.

Epilogue

The machine shop business has grown into a larger shop headed by Bart and Kurt. They had two employees already, and I have become the third. My position is what might be called gopher, or errand boy, but I am also learning how to operate the automatic milling machines. They cannot pay me what I had been making at Fairfield, but believe me, I have not been complaining.

The shop is located in what used to be a strip mall, now completely empty of retail businesses. The space was a clothing store years ago, and the entire front is a set of glass windows facing the street. In the back of the shop are the milling machines, in the front the lathes. Most of the machines are computer controlled. Because of several break-ins, the Kozys have spent a lot of time and effort in installing alarms and other security systems in the building. The strongest protection for the shop, however, is that both of the brothers live in the space upstairs, sleeping on mattresses on the floor.

A quick mental inventory of the machines and tools owned by the shop indicates that the company is worth more than eleven million dollars. The only debt the company has ever had was a quarter million dollar loan from Uncle Milt that was paid off in full two years ago.

In the very back of the shop is a large roll-up door where I handle receiving and shipping during most of the day. Sometimes, it can be discouraging to think that I had invested those years in school getting the chemical engineering degree. Most of the people that I meet in the shop do not know I have this unused diploma and certainly none of them care.

Every day at lunchtime, Mrs. Kozy comes by with sandwiches and drinks for her two youngest sons, and for me. Midge knows that I like pickles and horseradish on my Rueben. There were many days of the talk being nothing other than polite conversation about the weather and the business climate. I had been hesitant to broach the subject of Bruce, her oldest son.

One day at lunch, Kurt brought up the idea that the machine shop could try to get work from the Fairfield Lab across the bay. "Bruce is a department head or something over there. He said that they need all sorts of machine work done all the time."

Bart chimed in: "They have an exclusive contract with Butterfield machining in San Leandro . . . there is no chance of us getting any work there unless we were to subcontract from Butterfield." I just bit my tongue, and thought about the millions of dollars of pure waste and fraud that Bruce oversaw. Just a few hundred thousand dollars sent our way would not even be noticed.

I then asked Midge if she could explain what happened to Bruce. "Midge, Bruce is not like the rest of us. He is evil." When I said that last word, I found myself getting tense. I thought maybe this is not the best thing to tell a mother about her son. Then, I continued anyway, since I had already let it out. "Do you think Bruce is demon possessed?"

"Demons are everywhere and they try to influence us all the time. Bruce is heavily under demon influence, but he is not demon possessed," she said.

"You see, Stan," she continued, "this world was created by God because of a wager that Jesus made against the dark angel Lucifer in eternity past. Lucifer, who has now become Satan, said that if humans were allowed free will, the entire human race would eventually join him in becoming evil."

Then she said, "Satan did not understand that he could not win this bet. Satan cannot comprehend goodness. Jesus knows that once there is any good in a person, then that person cannot choose pure evil again. *God rigged the game against evil from the very beginning. That is why Satan will lose.*"

"It's a game?" I said, incredulously. She used the same word Bruce had used on the telephone from Russia. It's a big game and we are just the players. Now I know that he game is rigged; thank God for that.

I found out later that the inspector general at Fairfield followed the trail of information provided by the program manager, Doug Sweeney, and it was a very rich trail indeed. The IG essentially occupied a wing of the first floor of the ATUM building for over a year, after which Don Getz took an early retirement. Getz received special recognition for Distinguished Scientific Service from the lab director. Bruce, who was unscathed by the investigation, was made the new division leader. He appointed Mark as his deputy. They are required to work under a new operating contract, which has disallowed much of the wanton fraud. Bruce claims, I hear, it is my fault that the new contract is so restrictive. According to him, I am nothing but a scoundrel, if not a traitor.

Karen is living in, of all places, Palo Alto, just a few miles from the Kozy house. DNA testing confirmed that, indeed, I was the father of her son, Sean. Child custody arrangements were made with the help of some good attorneys. I visit with Sean every other weekend. Karen seems more relaxed and at ease. She is using some sort of medication for what she calls her "bi-polar disorder." She is always trying to get me to go to Alcoholics Anonymous meetings with her, but so far, she has had no luck. Those AA people are all, like her, totally crazy.

After returning Sean back to Karen after a recent visit, I stood there in the machine shop and wondered what had made Karen come after me so relentlessly those days a few years back in Berkeley. After all, there was that brutal ugly scar on the side of my face. No woman could possibly want a man who looked as horrible as I did.

I walked across the shop floor, into the bathroom to look in the mirror. The scar was gone. In fact, until that moment, I had forgotten about the scar completely.

About the Author

William Sailor lives in Rio Rancho, New Mexico. He holds a PhD in Engineering Science from UC Berkeley. He retired from Los Alamos National Laboratory in 2010 after 23 years of service. This book is his first fictional work.

CPSIA information can be obtained at www.ICGtesting.com
Printed in the USA
LVOW10s2134140913

352459LV00001B/1/P